Get the eBook FREE!

(PDF, ePub, Kindle, and liveBook all included)

We believe that once you buy a book from us, you should be able to read it in any format we have available. To get electronic versions of this book at no additional cost to you, purchase and then register this book at the Manning website.

Go to https://www.manning.com/freebook and follow the instructions to complete your pBook registration.

That's it!
Thanks from Manning!

Build Your Own Robot

USING PYTHON, CRICKIT, AND RASPBERRY PI

MARWAN ALSABBAGH

MANNING
SHELTER ISLAND

For online information and ordering of this and other Manning books, please visit www.manning.com. The publisher offers discounts on this book when ordered in quantity. For more information, please contact

Special Sales Department
Manning Publications Co.
20 Baldwin Road
PO Box 761
Shelter Island, NY 11964
Email: orders@manning.com

Manning Publications Co.
20 Baldwin Road
PO Box 761
Shelter Island, NY 11964

Development editor: Karen Miller
Technical editor: Alexander Ryker
Review editor: Adriana Sabo and Dunja Nikitović
Production editor: Andy Marinkovich
Copy editor: Lana Todorovic-Arndt
Proofreader: Melody Dolab
Technical proofreader: Alexander Ryker
Typesetter and cover designer: Marija Tudor

ISBN 9781633438453
Printed in the United States of America

To my loving family.

brief contents

contents

preface

I love learning how to create things with computers, whether these are video games or web applications. When I became a father, I also started creating fun crafting projects with my daughters. As they grew, our projects became increasingly ambitious, until we started building robots together. One of these fun robotics projects was creating a robot that could drive around and fetch books from a high bookshelf my daughter couldn't reach. The books would come flying off the shelf at high speed, and she would have to either catch or dodge them.

I presented the code and process of creating these projects at different Python conferences, much to the amusement of the audience. This is what opened up the idea of writing a book on creating robotics from scratch for people like me with no prior background in robotics. All you would need is to be familiar with Python and have a passion for learning new and fun ways of building interesting projects with the language.

There were a number of qualities that I wanted to make sure this book had. These were based on what I felt worked best when learning a new topic, and I also wanted to address certain shortcomings of some of the learning material already out there:

- *Offer results as early as possible.* From the second chapter, you'll be able to write and run Python code that will move motors and interact with sensors.
- *Be hands-on.* The book takes a very practical approach to building a robot from scratch.
- *Be accessible.* No special robotics, soldering, or electronics knowledge is required to create the projects in the book. Any Python developer can get started building these robots.

- *Offer many projects.* There are nine different projects covered in the book, and after the initial getting-started chapter, each chapter ends with a newly completed project.

With these guiding principles, the structure and content of the book took form. I hope you enjoy reading and running the projects as much as I did creating them.

acknowledgments

I'd like to thank Karen Miller, my editor at Manning. This has been my first book with Manning, and you have been incredibly patient and supportive throughout the process. Thank you for all the support and valuable feedback! I'd also like to extend my thanks to the technical editor, Alexander Ryker, for the excellent feedback and testing that he performed on the code in the book. A special thanks to the teams that worked on the promotion and production of the book. You helped me dig deep to understand why I am so passionate about this book and to share that passion with my readers.

I'd also like to thank the reviewers who gave such detailed and excellent feedback. To Alain Couniot, Alena Coons, Alex Lucas, Amitabh Cheekoth, Ben McNamara, Chad Yantorno, Cosimo Attanasi, Darrin Bishop, Erico Lendzian, Fernando Bernardino, James Black, James Matlock, Jeremy Chen, Jesús Juárez, Jon Choate, Jonathan Reeves, Julien Pohie, Keith Kim, Marc Taylor, Marcus Geselle, Martin Dehnert, Mohana Krishna BG, Patrice Maldague, Patrick Regan, and Richard Tobias, your comments were varied and helped me see the book from another angle. The changes and improvements we made wouldn't have been possible without your valuable observations.

about this book

Build Your Own Robot is a DIY guide for bringing your first Python-based robots to life. Starting with the basics, you'll teach your new friend how to spin, move around, and find its way. You'll then quickly progress to controlling your robot remotely using your phone, computer, or joystick. You'll even set up a camera to broadcast what it sees right to your computer screen. Clever computer vision tricks will get your robot tracking faces, looking for QR codes, and maybe even fetching some snacks.

Who should read this book?

This book is geared toward software developers, and the reader should be familiar with Python. No prior knowledge or experience in robotics or electrical engineering is required. All the hardware assembly in the book can be performed with simple tools such as a screwdriver. No special tools or skills such as soldering are required for any of the wiring or assembly of the robots. This book is very well suited for

- Python developers
- Robot enthusiasts
- University students

How this book is organized: A roadmap

The book has 11 chapters. Both beginner and experienced Python developers can learn from the software techniques used in the book to bring the robotics projects to life:

- Chapter 1 explains why robots are so amazing and why they have so much potential. It also discusses the building blocks of the robots and the approach that will be taken to building the robots in the book.

- Chapter 2 covers the initial steps to setting up a robot and getting started with robotic projects. Creating software to control the DC motors and change Neopixel colors based on touch sensors will also be covered.
- Chapter 3 explores the topic of making your robot move around. You will learn how to control the DC motors to make the robot move forward and backward, as well as turn left and right. All these different movement functions will be put into a library so that you can reuse the code in later chapters.
- Chapter 4 covers the basics of creating an interactive custom shell in Python so that you can create a robot shell. The shell will support commands to perform different robot movements and have a command history and the ability to execute custom shell scripts to make the robot perform a sequence of movements.
- Chapter 5 discusses the topic of creating software to control robots remotely. The SSH and HTTP network protocols will each be used as popular options for remote control.
- Chapter 6 discusses how to create robot web apps that can be used to control the robot with a phone or computer web browser. Topics such as measuring web application performance using web browser tools will also be covered.
- Chapter 7 covers the topic of controlling robots with joysticks. Different approaches to reading and responding to joystick events in Python will be examined. Then, an application to respond to joystick events with different robot movements over the network will be created.
- Chapter 8 discusses how to control a set of servo motors to perform movements in a pan and tilt directions with a keyboard. A camera mounted to the robot servo motors can then be controlled to display a live video stream and take snapshots.
- Chapter 9 covers the topic of building a robot that moves the camera in the direction of a detected face so that the camera will follow it. Machine learning will be used to perform face detection and move the camera based on the location of the detected face in the captured video frame.
- Chapter 10 sets out to build a robot that can move around in search of a matching QR code in its environment. Techniques for generating and detecting QR codes in Python are covered, as well as how to build an application that will drive the robot around until it finds a matching QR code.
- Chapter 11 discusses how to build a snack-pushing robot that reads a list of snacks from a CSV file. A desired snack can then be selected from a web application to make the robot take action and drive to the selected snack and push it.

About the code

This book contains many examples of source code, both in numbered listings and in line with normal text. In both cases, source code is formatted in a `fixed-width font like this` to separate it from ordinary text.

You can get executable code snippets from the liveBook (online) version of this book at https://livebook.manning.com/book/build-your-own-robot. The complete code for the examples in the book is available for download from the Manning website at www.manning.com and from GitHub at https://github.com/marwano/robo.

liveBook discussion forum

Purchase of *Build Your Own Robot* includes free access to liveBook, Manning's online reading platform. Using liveBook's exclusive discussion features, you can attach comments to the book globally or to specific sections or paragraphs. It's a snap to make notes for yourself, ask and answer technical questions, and receive help from the author and other users. To access the forum, go to https://livebook.manning.com/book/build-your-own-robot/discussion. You can also learn more about Manning's forums and the rules of conduct at https://livebook.manning.com/discussion.

Manning's commitment to our readers is to provide a venue where a meaningful dialogue between individual readers and between readers and the author can take place. It is not a commitment to any specific amount of participation on the part of the author, whose contribution to the forum remains voluntary (and unpaid). We suggest you try asking the author some challenging questions lest their interest stray! The forum and the archives of previous discussions will be accessible from the publisher's website for as long as the book is in print.

Software/Hardware requirements

The hardware purchasing guide in appendix A covers the hardware requirements for the book projects. It shows which hardware is needed for different chapters. It also has some recommendations for optional purchases that can help improve robot projects. Appendix B provides detailed instructions for the installation and configuration of the Raspberry Pi and the related software requirements. For details on assembling and configuring the robot hardware, check the robot assembly guide found in appendix C. Finally, appendix D provides a mechanism to mock the robotic hardware and run all the code in the book on any laptop or desktop computer.

about the author

MARWAN ALSABBAGH is a seasoned software developer. He studied mathematics and computer science at McGill University and is passionate about teaching and learning by building projects using Python, with a focus on microcontrollers and robotics.

ABOUT THE TECHNICAL EDITOR

ALEX RYKER is a consulting systems engineer working in the industrial automation sector. He majored in computer science at Purdue University and has co-authored research papers in the fields of computer security and robotics.

about the cover illustration

The figure on the cover of *Build Your Own Robot,* titled "*L'Etalagiste,*" is taken from a book by Louis Curmer, published in 1841. Each illustration is finely drawn and colored by hand.

In those days, it was easy to identify where people lived and what their trade or station in life was just by their dress. Manning celebrates the inventiveness and initiative of the computer business with book covers based on the rich diversity of regional culture centuries ago, brought back to life by pictures from collections such as this one.

1 What is a robot?

> **This chapter covers**
> - What robots are made of
> - Why robots have so much potential
> - Hardware and software components used by robots

Recently, the field of robotics has grown tremendously, with robots for consumer usage being increasingly replaced by those for industrial application. The hardware and software behind these robots have also become more accessible, and for this reason, these are exciting times to learn about robotics. This book employs the power of the Python programming language to bring a wide variety of robotics projects to life using software and hardware that embrace the open source philosophy. By the end of the book, you will learn how to build nine different robotics projects.

1.1 Why robots are amazing

Computers have changed the lives of every human on this planet in ways their creators could not have even imagined decades ago. Robot technology gives those computers the arms and wheels to move around and achieve things beyond our imagination. In many ways, robots are the future. This book allows you to build robots from the ground up and see them come to life in the real world. The code you write will make your computer drive around and knock over items of your choosing. Figure 1.1 provides a visual illustration of different ways in which we can use the power of robots.

Figure 1.1 Power of robots: robots use different hardware to move and see the world around them.

Robots can do many things:

- *Robots move.* They have wheels and are not afraid to use them. This mobility gives rise to having them perform all sorts of tasks, from transporting packages in a warehouse to transporting people in self-driving cars.
- *Robots can see.* With a camera on board and powerful computer vision software, robots can see the world around them. They can see our faces, detect them, and react to them by moving their camera in our direction using motors.
- *Robots can find.* Using QR code stickers on objects in the real world, we can send our robot off to find and interact with objects of our choosing. Robots use their cameras and QR detection software to find what they are looking for and perform the task at hand once they have arrived at their destination.
- *Robots can be controlled.* Human-operated robots let us perform all kinds of jobs that were not possible before (e.g., controlling robots in areas too hazardous for humans to enter or performing medical procedures too delicate for human hands). In this book, we will use keyboards, mice, and joystick controllers to control robots from computer screens and mobile phones. Each device has its benefits and tradeoffs, which will be explored in each implementation.
- *Robots can be commanded remotely.* One of the most powerful computer innovations of our times is the TCP/IP internet communication protocols. Using HTTP and SSH, which are built on top of TCP, we will implement different ways of remotely communicating with and controlling our robots. Whether the human operator is in the same room or miles away, these robots will be able to be controlled.

1.2 Our approach to making robots

This book takes a very hands-on approach to learning how to build robots and write software to control and interact with them in Python. By the end of the next chapter, you will assemble and connect enough components of the robot to write and run your first Python script. This script will read the state of the onboard touch sensor as input and turn on the DC motor when it detects a touch event. Each remaining chapter ends with a fully functional project that is either directly interacting with the robot or is a stepping stone to adding more functionality to other robots later in the book. Each of these project-based chapters will introduce the project at hand and then guide the reader step by step as the solution gets constructed.

1.2.1 Learning from failures

The most direct approach will often fail because of hardware constraints in processing power or inherent limitations in a certain technology. These situations will be used as learning opportunities to see how to overcome these constraints by using more sophisticated approaches or by optimizing the implementation to gain major performance boosts to work under constrained hardware requirements. The reader will be brought along on the ride of diagnosing, precise measuring, and benchmarking these problems and their associated solutions. The reader will be able to use these skills to predict, diagnose, and debug similar problems in their own projects. Many of these performance problems and solutions are not limited to working on robots and apply to many different computer vision, networking, and computation projects.

1.2.2 What will you gain?

By reading this book, you will be exposed to a wide variety of hardware and software challenges. Many of the Python libraries in this book are widely used and are suitable for projects involving robots and beyond. Many problem-solving techniques for dealing with hardware and software problems are also covered in the book, which will be beneficial to the reader.

The field of robotics is a very broad field with a wide variety of applications. In terms of the software and hardware covered in this book, a specific set of hardware and software will be used.

1.2.3 Prototyping

Using the hardware and software covered in this book, there is a wide variety of prototyping projects that can be completed. Because of the inexpensive but powerful hardware in use, many ideas can be tested by building a prototype to test drive the design or approach before committing to more involved or expensive production hardware.

1.2.4 Teaching

The projects in the book can be used to teach students in a classroom setting or even for self-study to gain real-life experience with building robots. The underlying technology of robots is the same, whether you create a small robot or a larger one for industrial use. By building these robots, you will also gain knowledge related to chassis design, motor and computer power needs, battery power capacity, weight, and portability. The lessons learned on a small scale are still very applicable to larger or production robots.

1.2.5 Production ready

As for the software, much of the software used in the book is directly suitable for large-scale production. Software such as Linux, Python, and the Tornado web framework are used in many mission-critical applications. As for the hardware, the Raspberry Pi is used in many end-user products, and the Raspberry Pi Compute Module is heavily used for industrial applications.

1.2.6 Limitations

There are inherent hardware limitations in the output and power capacity of the motors used in this book, which will directly limit the types of projects that can be executed with this hardware. The computing power of the Raspberry Pi is also inherently more limited, as it is designed to be small and lightweight with low power consumption.

1.3 What are robots made of?

Robots frequently have processors, memory, and storage like any laptop or desktop computer. But what sets them apart are their motors and sensors, which let them do things no regular computer could. Add some powerful software to the mix, and they will be able to perform all sorts of feats.

1.3.1 The robot building blocks

It is valuable to build a mental model of the hardware and software used in this book. It will help you understand which part of the software and hardware stack we are operating in. Linux and Python are an ecosystem of incredibly powerful software modules and libraries that are well tested, versatile, and well documented. We will take advantage of this and combine many different modules to achieve the functionality we need.

1.3.2 Servos and DC motors on a Raspberry Pi

DC motors turn the wheels in robots and help them drive around. Whether they drive in all different directions or along a fixed path or track, it is DC motors that make it possible. *Servos* are more sophisticated motors with built-in sensors that allow precise movements. They can be controlled to turn to specific angles. Robotic arms are often built from several servo motors. The projects in the book will cover heavily both servo and DC motors and will control them from Python scripts to perform a wide variety of tasks. The robots in the book will have a small but powerful *single-board computer* (SBC), called the Raspberry Pi, at the heart of their operation.

1.3.3 Hardware stack

Figure 1.2 shows different hardware components that will be used for different projects in this book.

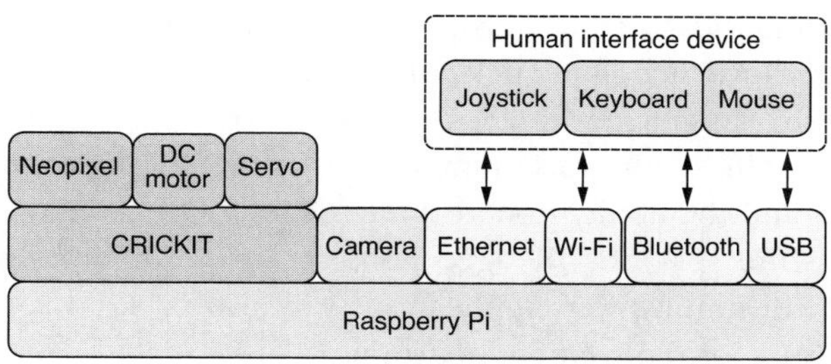

Figure 1.2 Hardware stack: the Raspberry Pi is the main computer, and the CRICKIT handles the motors.

Each hardware component is described as follows:

- *Raspberry Pi*—An SBC that has a CPU, RAM, and several different input/output interfaces.
- *CRICKIT*—The Adafruit CRICKIT HAT for Raspberry Pi is an add-on board that attaches to the general-purpose input/output (GPIO) connector of the Pi. It powers and controls the servo and DC motors.
- *Neopixel*—The CRICKIT HAT comes with an RGB LED. Its color and brightness can be changed in the software.
- *DC motor*—Multiple DC motors can be connected to the CRICKIT HAT. They will provide power to the left and right wheels to drive the robot.
- *Servo*—Multiple servo motors can be connected to the CRICKIT HAT. They can be used for the camera pan and tilt of the servo motors.
- *Camera*—The Raspberry Pi Camera Module is created by the Raspberry Pi Foundation and is used for computer vision.
- *Ethernet*—The gigabit Ethernet port is built into the Pi board and provides high-speed, low-latency reliable network communication.
- *Wi-Fi*—Provides dual-band 2.4 GHz and 5 GHz wireless LAN.
- *Bluetooth*—Bluetooth 5.0, BLE
- *USB*—The keyboard, mouse, and joysticks can be connected directly using USB.

Different projects will have different sets of required hardware components. Each chapter will present the hardware components that will be used for a particular project. The hardware-purchasing guide in appendix A provides good advice on the specific models of products needed for the projects, as well as details on online resellers.

1.3.4 Python and Linux

The robots will be running their software on top of the Linux operating system. Linux is a feature-rich and versatile operating system that powers robots and supercomputers alike. This opens the door to a wide variety of mature and well-tested software and features that will be used throughout the book—everything from live video processing from the camera feed to using the Bluetooth protocol to wirelessly control the robot's movements using highly sensitive analog joystick controllers.

Python is a very expressive language with a wide selection of mature and powerful software libraries. We will use these libraries to incorporate rich functionality into the robot-like computer vision and the ability to detect and decode QR codes so that the robot can search for and find specific QR-tagged objects in its surroundings. The libraries for creating and consuming web applications will also be used so that the robot can be controlled over a local or remote network.

1.3.5 Software stack

Figure 1.3 shows different pieces of software that will be used throughout the projects in the book.

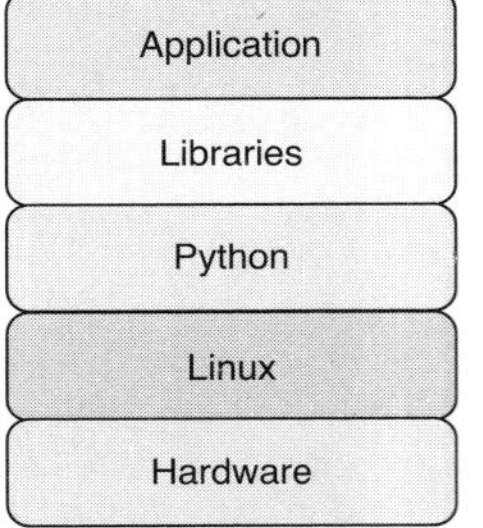

Figure 1.3 Software stack: Linux manages the hardware and runs the Python interpreter.

The software components of a robot are as follows:

- *Linux*—The Raspberry Pi OS Linux distribution will be used as the operating system.
- *Python*—The Python interpreter will run as an executable on Linux and execute the robot Python applications.
- *Libraries*—These are a variety of Python libraries that will be used and incorporated into the robot projects, all running within the Python interpreter. The Tornado web framework will be used to build web applications to control the robot. The OpenCV computer vision library will be used for face detection and QR code decoding. The Adafruit CRICKIT library will be used to control servo and DC motors.
- *Application*—The code driving the robot projects in this book will be operating at this level.

Summary

- With a camera on board and powerful computer vision software, robots can see the world around them.
- By scanning QR code stickers on objects in the real world, robots can find and interact with objects in their environment.
- Human-operated robots let us perform all kinds of previously impossible jobs, whether these imply controlling robots in areas too hazardous for humans to enter or performing medical procedures too delicate for human hands.

- Using internet protocols such as HTTP and SSH, we will implement different ways of remotely communicating with and controlling our robots.
- Many different types of prototyping projects can be executed using the hardware and software covered in this book.
- Many of the Python libraries in this book are widely used and can be employed for projects involving robots and beyond.
- Robots frequently have processors, memory, and storage like any laptop or desktop computer.
- Robots also often run Linux, a feature-rich and versatile operating system that powers robots and supercomputers alike.
- DC motors are what turn the wheels in robots and help them drive around.

2 Getting started

This chapter covers

- Assembling and configuring the core hardware and software of the robot
- Controlling the Neopixel color and brightness
- Reading sensor data from the four onboard touch sensors
- Controlling DC motors with Python
- Creating your first Python robotics program that interacts with sensors and motors

In this chapter, you will learn how to connect and configure the main hardware and software components used for the robots in this book. Once the hardware and software are set up, we'll get straight into interacting with the hardware using Python by reading sensor data from the onboard touch sensors. Then, you will learn how to control the Neopixel lights and a DC motor. Finally, all these different hardware components and Python scripts will come together to create a robotics program that controls the Neopixel and DC motors based on touch sensor input.

2.1 Introducing our robotic hardware

Figure 2.1 shows the hardware stack discussed in the previous chapter, with the specific components used in this chapter highlighted in the boxes with darker text. The components with grayed-out text will be used in later chapters.

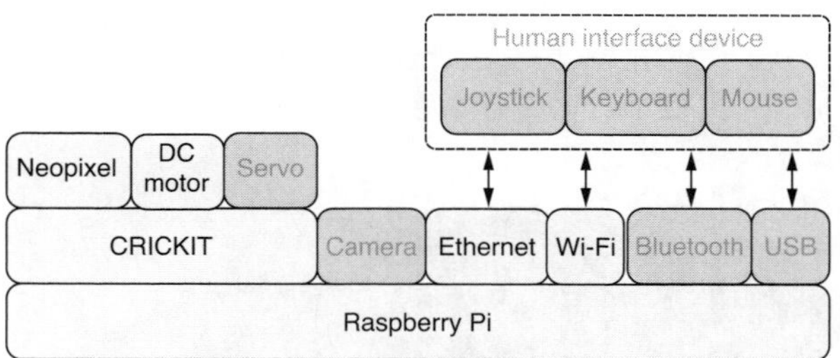

Figure 2.1 Hardware stack: the Raspberry Pi will handle network communication using Ethernet and Wi-Fi.

The Linux operating system will be installed on the Raspberry Pi. The Ethernet and Wi-Fi hardware components will be used to connect the machine to the network and allow other computers on the network to connect to it. The CRICKIT HAT will then be connected to the Raspberry Pi and used to control the Neopixel and attached motor.

Make sure to check the hardware purchasing guide found in appendix A before buying the hardware needed in this chapter. The appendix shows which hardware is needed for the different chapters and also has some recommendations for optional purchases that can improve robot projects. Appendix B provides detailed instructions for the installation and configuration of the Raspberry Pi and Adafruit CRICKIT HAT. It is also worth noting that appendix D provides a mechanism to mock the robotic hardware and run any code in the book on any laptop or desktop computer.

2.1.1 Raspberry Pi

The Raspberry Pi is a small single-board computer created by the Raspberry Pi Foundation (https://raspberrypi.org). The Raspberry Pi 4 with 2 GB or more of RAM is the recommended model to use for the projects. Figure 2.2 shows a photo of the Raspberry Pi 4 for reference. The attributes of these computers that make them an ideal choice for robotics projects are the following:

- Their small size and lightweight structure are important for mobile robots.
- Running Linux and Python opens the door to a wide array of powerful software to build robot projects on.
- Versatile robot chassis kits compatible with the Raspberry Pi allow different configurations of board, motor, and battery.
- Powerful CPU and memory make intensive applications such as real-time computer vision and machine learning possible.

- Good camera support allows the robots to see their environment.
- They have built-in and flexible connectivity options such as Ethernet, Wi-Fi, Bluetooth, and USB.
- A general-purpose input/output (GPIO) connector provides a powerful mechanism to add hardware features to the board utilized by the Adafruit CRICKIT.

Figure 2.2 Raspberry Pi: the main hardware interfaces on the board are labeled.

2.1.2 Adafruit CRICKIT HAT

The Adafruit CRICKIT HAT is a hardware add-on for the Raspberry Pi created by Adafruit Industries (https://adafruit.com; figure 2.3). The CRICKIT HAT connects to the GPIO connector on the Raspberry Pi and provides the following features used by the projects in the book:

- Up to two bi-directional DC motors can be connected, powered, and controlled.
- Up to four servo motors can be connected, powered, and controlled.
- Four capacitive touch input sensors are built into the board.

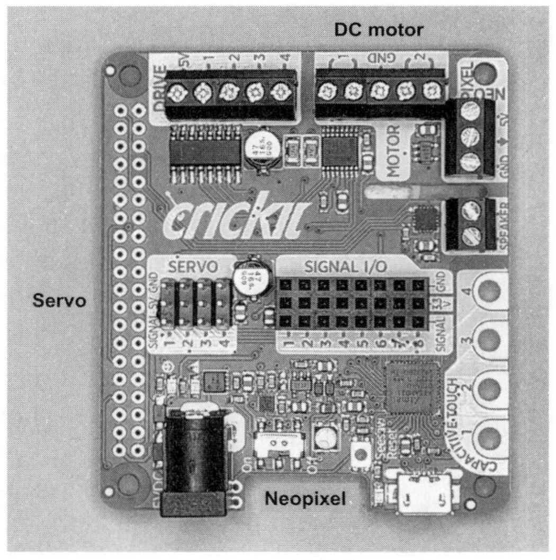

Figure 2.3 Adafruit CRICKIT HAT: the DC motors and servo motors are connected to this board.

- A Neopixel RGB LED is built into the board.
- Python support is provided by the Adafruit Python library to control and interact with the motors, capacitive touch, and Neopixels.

2.2 Configuring the software for our robots

Figure 2.4 shows the software stack presented in the previous chapter. Details of the specific software used in this chapter are described in the following text.

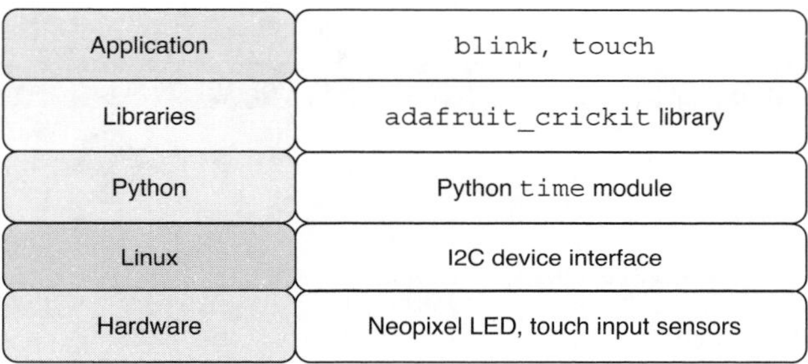

Figure 2.4 Software stack: this chapter will cover the installation and configuration of Linux and Python.

Once Linux is installed, Python will be configured to have a dedicated virtual environment where Python libraries can be installed. The Adafruit CRICKIT library will be installed, which will then be used to run the Python code to interact with the CRICKIT hardware components such as the Neopixel LED and the touch input sensors. The `blink` and `touch` applications will then be created using Python and the Adafruit libraries to communicate with the CRICKIT board using the I2C device interface in Linux. The Python `time` module will be used to control the time duration of different actions. Before continuing with the chapter, make sure to follow the installation and configuration instructions in appendix B for the Raspberry Pi and the Adafruit CRICKIT HAT.

2.3 Changing Neopixel colors

The CRICKIT library offers several different ways to interact with the Neopixel LED. We can explore these options in a REPL (read–evaluate–print loop) session. Check appendix B for help activating the Python virtual environment and opening a REPL session. The Neopixel can be quite bright, so we will lower the brightness to 1% and then set the color to blue:

```
>>> from adafruit_crickit import crickit
>>> crickit.onboard_pixel.brightness = 0.01
>>> crickit.onboard_pixel.fill(0x0000FF)
```

So far, we have set the color using the RGB hexadecimal color code. It would be nice if we could set colors using human-readable color names. This functionality isn't directly available through the CRICKIT library, but we can create a simple dictionary to store and look up common color names:

```
>>> RGB = dict(red=0xFF0000, green=0x00FF00, blue=0x0000FF)
>>> crickit.onboard_pixel.fill(RGB['red'])
```

We can now create a simple script to continually loop through each color name and set the color. This code will create a multicolor blinking effect with the LED. During each loop, the script will print out the color name, set the color, and pause for 0.1 seconds before setting the next color. Save the following script in a file called `blink.py` on the Pi.

Listing 2.1 `blink.py`: Creating a multicolor blinking effect with the LED

```
#!/usr/bin/env python3
import time
from adafruit_crickit import crickit

RGB = dict(red=0xFF0000, green=0x00FF00, blue=0x0000FF)

crickit.onboard_pixel.brightness = 0.01
while True:
    for name in RGB:
        print(name)
        crickit.onboard_pixel.fill(RGB[name])
        time.sleep(0.1)
```

The file can be given execute permission by running

```
$ chmod a+x blink.py
```

Then run the Python script:

```
$ ./blink.py
```

The reason the script can be executed directly is that the first line is using a Unix feature called shebang, which tells Linux that the script should be executed through the Python interpreter `python3`. Make sure to activate the Python virtual environment before running the script, as shown in appendix B. We can exit the script by pressing Ctrl+C, which will force the script to exit. When saving the script on the Pi, place it in the `/home/robo/bin/` directory, which can also be accessed as `~/bin`. This is a standard location on Linux systems to place user scripts such as these, and it will be the convention followed in the book. The `blink.py` file and all the code for the projects in this book can be found on GitHub (https://github.com/marwano/robo).

Going deeper: The I2C communication protocol

The CRICKIT HAT has its own microcontroller on the board and uses the I2C communication protocol to enable communication between the Raspberry Pi and its microcontroller. This is all taken care of by the Python Adafruit CRICKIT library. Being very powerful and flexible, I2C protocol is a popular choice for communication between integrated chips. The SparkFun website has a great guide (https://learn.sparkfun.com/tutorials/i2c) on I2C. It can be interesting and useful to learn what is going on under the hood with these low-level hardware protocols.

The Adafruit site has a good hands-on guide (http://mng.bz/g7mV) to communicating with I2C devices in Python. We can use this guide to do some basic interaction with I2C on the Pi with the CRICKIT HAT. Let's first open a REPL and import the `board` module:

```
>>> import board
```

We can now create an `I2C` object to scan for the CRICKIT HAT:

```
>>> i2c = board.I2C()
```

We can now scan for I2C devices and save the result in `devices`. We can see from the results that one device was found:

```
>>> devices = i2c.scan()
>>> devices
[73]
```

From appendix B, we know that the I2C address for the CRICKIT HAT is expected to be `0x49`. We can confirm the device we found is the CRICKIT HAT by calculating its hexadecimal with the following line:

```
>>> hex(devices[0])
'0x49'
```

The I2C protocol is a powerful protocol that can support up to 1,008 peripheral devices connected on just two wires.

2.4 Checking the touch sensor state

There are four capacitive touch input sensors on the CRICKIT. Figure 2.5 shows their close-up view. From Python, each sensor can be checked individually to see whether it is currently detecting a touch event. Without touching the touch sensor, run the following code in a REPL session:

```
>>> from adafruit_crickit import crickit
>>> crickit.touch_1.value
False
```

Now run the last line again while touching the first touch sensor:

```
>>> crickit.touch_1.value
True
```

When the `value` attribute is accessed, the CRICKIT library checks the touch sensor state and returns a Boolean value of either `True` or `False` depending on the sensor data.

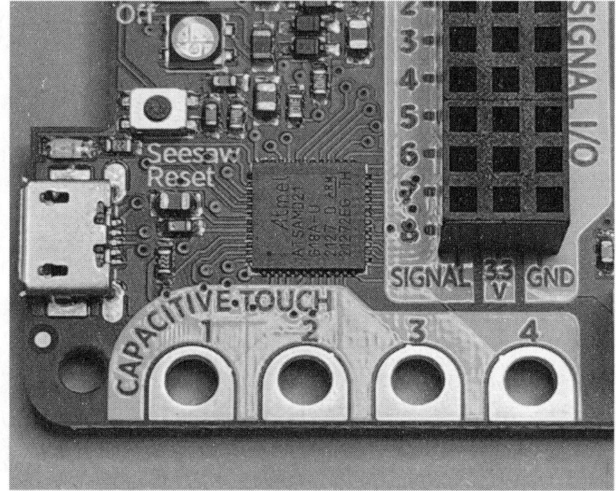

Figure 2.5 Capacitive touch sensors: these sensors can detect touch events.

2.5 Controlling DC motors

Connect the two wires of a DC motor into the DC motor connector port 1. Figure 2.6 shows the location of these motor connections on the CRICKIT. The two wires can be connected either way into the CRICKIT motor port; it won't cause any problems. Make sure to use the M/M extension jumper wires mentioned in appendix A, as this will ensure the male and female ends match. Once connected, run the following lines in a REPL session:

```
>>> from adafruit_crickit import crickit
>>> crickit.dc_motor_1.throttle = 1
```

The DC motor will now run at full speed. To stop the motor running, use

```
>>> crickit.dc_motor_1.throttle = 0
```

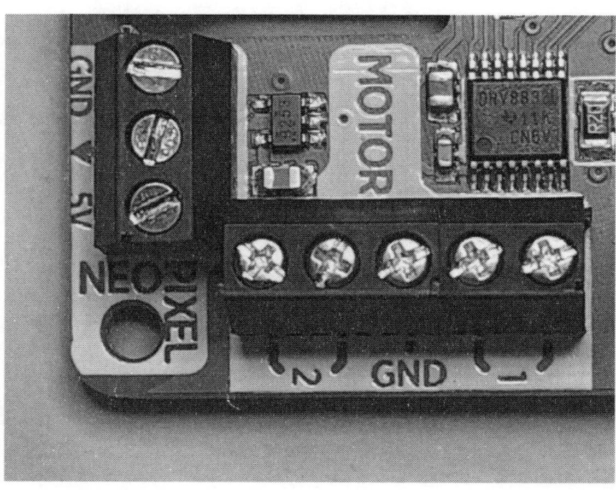

Figure 2.6 DC motor connections: the connection points for DC motors are screwed in place.

2.6 Controlling motors with the touch sensor

We can take what we have learned so far and combine these different parts of the CRICKIT library to make an application that continually checks the touch sensor and starts or stops the motor based on whether the sensor is being touched. We will also change the LED color whenever the motor is started or stopped. Let's start constructing this application piece by piece.

First, we'll import the CRICKIT library to control the motors and the `time` library to pause between checking for touch events:

```
import time
from adafruit_crickit import crickit
```

Next, we'll define `RGB` so that we can set colors by name and save how long we wait between checking touch events in a setting called `POLL_DELAY`. The poll delay value is set at 0.1 seconds, which is fast enough to make the experience of touching the sensor and starting the motor responsive:

```
RGB = dict(red=0xFF0000, green=0x00FF00, blue=0x0000FF)
POLL_DELAY = 0.1
```

Before starting the main loop of the program, we set the brightness of the LED:

```
crickit.onboard_pixel.brightness = 0.01
```

The remainder of the program runs in this infinite loop:

```
while True:
```

In the first line of the loop, we check whether the sensor is being touched and set the `throttle` variable accordingly. In Python, this syntax is called a *conditional expression*:

```
throttle = 1 if crickit.touch_1.value else 0
```

We take the same approach to set the LED to red when the motor is on and blue when it is off using the `color` variable:

```
color = RGB['red'] if crickit.touch_1.value else RGB['blue']
```

After calculating the `throttle` and `color` values, we apply them to the motor and LED:

```
crickit.onboard_pixel.fill(color)
crickit.dc_motor_1.throttle = throttle
```

Finally, we sleep for `POLL_DELAY` seconds before starting the next loop iteration:

```
time.sleep(POLL_DELAY)
```

The full application can be saved as `touch.py` on the Pi and then executed.

Listing 2.2 `touch.py`: Starting the motor when the touch sensor is pressed

```
#!/usr/bin/env python3
import time
from adafruit_crickit import crickit

RGB = dict(red=0xFF0000, green=0x00FF00, blue=0x0000FF)
POLL_DELAY = 0.1

crickit.onboard_pixel.brightness = 0.01
while True:
    throttle = 1 if crickit.touch_1.value else 0
    color = RGB['red'] if crickit.touch_1.value else RGB['blue']
    crickit.onboard_pixel.fill(color)
    crickit.dc_motor_1.throttle = throttle
    time.sleep(POLL_DELAY)
```

When you run the script, you will see that the motor is initially not moving, and the LED color will be blue. If you press the touch sensor, the LED color will change to red, and the motor will start moving at full speed. If you stop touching the sensor, the LED will return to the initial blue color, and the motor will come to a full stop. Figure 2.7 maps the touch sensor state to the color of the LED and the motor running state.

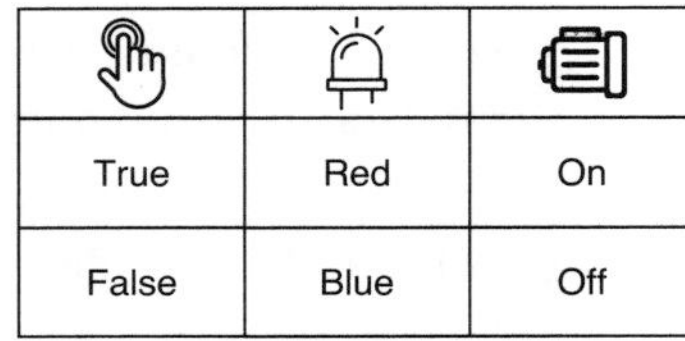

Figure 2.7 Touch state diagram: the LED and motor state changes in reaction to touch events.

Summary

- The Raspberry Pi is a single-board computer created by the Raspberry Pi Foundation.
- The Adafruit CRICKIT HAT is a hardware add-on for the Raspberry Pi created by Adafruit Industries.
- The CRICKIT HAT, once connected to the Raspberry Pi, can be used to control the Neopixel and attached motors.
- The Adafruit Python CRICKIT library can be used to run Python code that interacts with the CRICKIT hardware components.
- The Raspberry Pi Imager is software that can be used to prepare the installation media (microSD card/USB flash drive) with the Raspberry Pi OS image.
- The CRICKIT HAT is connected to the Raspberry Pi using the GPIO connector.
- Neopixel colors can be changed in Python using the RGB hexadecimal color codes.
- When checking the touch sensor state, the CRICKIT library returns a Boolean value of either `True` or `False`, depending on the touch sensor state.
- DC motors can be turned on and off by setting the throttle attribute to `1` or `0`.
- A connected motor can be turned on and off based on touch events using a loop that continually polls for touch events and sets the throttle for the motor accordingly.

3 Driving the robot

This chapter covers

- Controlling DC motors to make robots move forward and backward
- Implementing software-configured motor power adjustments
- Turning the robot left and right
- Spinning the robot in place
- Refactoring the code using the `functools` library

This chapter will teach you how to move the robot in different directions using the Python code to control the power given to DC motors. The left and right wheels of the robot each have a dedicated motor attached to them. Controlling the motors makes a whole array of movements possible. Python functions will be generated for each of the main movement operations to create an easy-to-use and readable set of methods to control the robot's movements. Once all these functions have been implemented, a number of refactoring techniques will be applied to simplify and consolidate the code base.

3.1 What's a robot chassis kit?

Robot chassis kits are a great way to build mobile robots. In the previous chapter, we saw how the Raspberry Pi provides computing power for the robot and how the CRICKIT HAT add-on controls connected motors. The chassis kit provides the body, motors, and wheels to get your robot moving around. There are many different robot chassis kits that can be used with the Raspberry Pi. The one recommended in this book is an inexpensive and flexible option that lets you build many different robot configurations. The kit comes with the following main parts:

- Two DC motors
- Two wheels
- One caster ball that acts as a third wheel
- Three aluminum frames and mounting hardware to create a three-layer robot

Figure 3.1 Chassis kit: the kit comes with three aluminum frames to support three-layer robots.

Figure 3.1 shows the main parts that come with the kit. Having three layers as opposed to a two-layer arrangement provides the expanded space and flexibility to install the Raspberry Pi and power hardware on different layers. The DC motors are kept on the bottom layer. These layers in the robot chassis can be seen in the images found in the robot assembly guide that is part of appendix C. Check the hardware-purchasing guide in appendix A for more details on the recommended chassis kit.

3.2 Hardware stack

Figure 3.2 shows the hardware stack discussed, and the specific components used in this chapter are highlighted. As the book progresses, more hardware components will be incorporated into the robot projects.

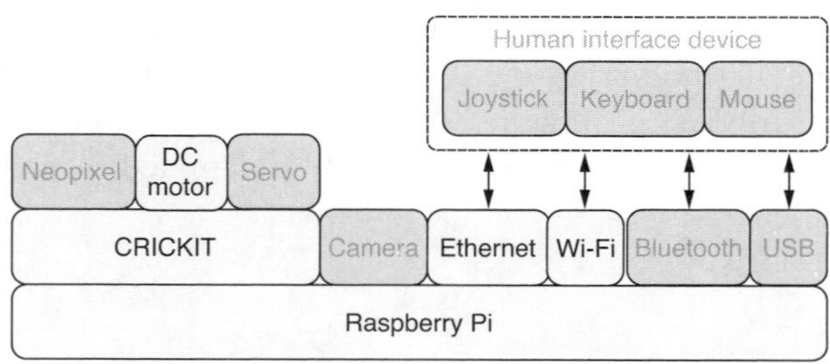

Figure 3.2 Hardware stack: the robot will move around using two DC motors.

As mentioned in the previous section, two DC motors will be used in this chapter. One will be attached to the left wheel and the other to the right wheel. The center wheel will use a caster ball that can turn smoothly in any direction and has no motor attached. The DC motors will be connected to the CRICKIT HAT, which will power and control them. Ethernet can be used to connect to the robot, but a Wi-Fi connection provides the ability for the robot to move around untethered.

For details on assembling and configuring the robot hardware, check the robot assembly guide in appendix C. It shows how to assemble the robot used in this and other chapters.

3.3 Software stack

Details of the specific software used in this chapter are illustrated in figure 3.3 and described in the text that follows. The application layer makes use of the libraries and the Python interpreter below it. The Python interpreter runs on Linux, which in turn runs on the hardware.

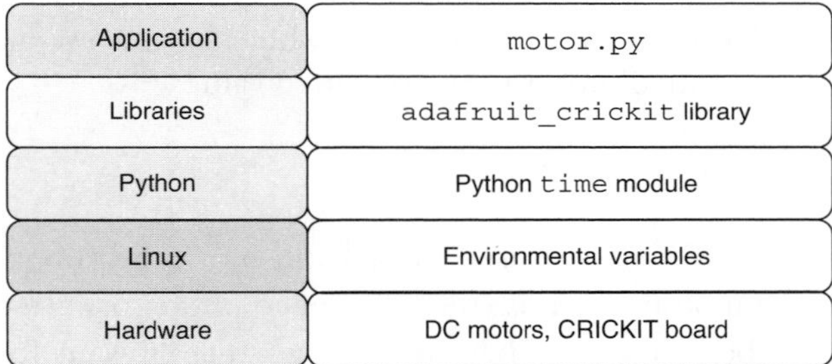

Figure 3.3 Software stack: this chapter covers how to control DC motors with Python.

Building on the knowledge from chapter 2, we will continue using the Python Adafruit CRICKIT library to interact with DC motors. In this chapter, we will control multiple DC motors using the same script. We will also learn how to control the motor direction and speed. On the Linux level, we will use environment variables such as `ROBO_DC_ADJUST_R` to pass configuration values to our Python scripts to set motor power adjustments. In this way, we don't have to hard-code configuration values directly into our Python code. The Python `time` module will be used to control the duration of the motor running during different movement operations. The `time` module is part of the Python standard library and provides a standard mechanism to pause script execution through the `sleep` function. Finally, we will bring all these pieces together to create the `motor.py` script and library at the end of the chapter.

3.4 Writing a move forward function

In this section, we will learn how to create a Python function that will make the robot move forward when called. We do this by turning on both the left and right DC motors at the same time, letting them run for a specific time, and then stopping them. This will make the robot move forward and come to a stop.

An important safety aspect when dealing with robots is having a mechanism to perform an emergency stop. The CRICKIT HAT has a hardware switch to turn the board on and off. We can use the switch as our emergency stop, as it will cut power to all connected motors. It has the added benefit that we can turn the board right back on and start our applications again without having to reboot the Raspberry Pi.

First, we'll import the CRICKIT library to control the motors and the `time` library to control how long we move forward:

```
from adafruit_crickit import crickit
import time
```

Next, we'll define `MOTOR_R` and `MOTOR_L`, which will map to the right and left motors. When wiring the robot, make sure to connect the right DC motor to motor connection 1 and the left DC motor to motor connection 2. All the code in the book will follow this convention:

```
MOTOR_R = crickit.dc_motor_1
MOTOR_L = crickit.dc_motor_2
```

Then, we define a helper function called `set_throttle`, which accepts two arguments and sets the throttle for the specified motor to the specified value:

```
def set_throttle(motor, value):
    motor.throttle = value
```

We can now define the `forward` function itself, which will move the robot forward. When called, it first sets both motors to run at the 90% speed in the forward direction by calling `set_throttle` on both motors. Then, by calling `sleep`, it waits for 0.2 seconds. Finally, calling `set_throttle` again on both motors ends the movement. In this way, calling the `forward` function will move the robot forward for 0.2 seconds and then cause it to come to a full stop:

```
def forward():
    set_throttle(MOTOR_R, 0.9)
    set_throttle(MOTOR_L, 0.9)
    time.sleep(0.2)
    set_throttle(MOTOR_R, 0)
    set_throttle(MOTOR_L, 0)
```

The full script can be saved as `forward.py` on the Pi and then executed.

Listing 3.1 `forward.py`: Making the motors move forward with a function

```
#!/usr/bin/env python3
from adafruit_crickit import crickit
import time

MOTOR_R = crickit.dc_motor_1
MOTOR_L = crickit.dc_motor_2
```

```
def set_throttle(motor, value):
    motor.throttle = value

def forward():
    set_throttle(MOTOR_R, 0.9)
    set_throttle(MOTOR_L, 0.9)
    time.sleep(0.2)
    set_throttle(MOTOR_R, 0)
    set_throttle(MOTOR_L, 0)

forward()
```

When running the script, both motors should move in the forward direction, making the whole robot move forward. If one of the wheels is moving in the backward direction instead of forward, just switch the connecting wires for that DC motor. The DC motor wires can be connected either way, and flipping the connected wires will also flip the direction of a positive throttle. The book follows the convention that a positive throttle results in a forward movement and a negative throttle results in a backward movement. If the wheel direction has not yet been confirmed, be mindful not to put your robot on the edge of a table, as it might drive in the wrong direction and fall off the table.

> **WARNING** The motor speed was set at 90% strength instead of a full 100% strength. This is done for power safety reasons. Using certain USB power banks to power the CRICKIT HAT and rapidly switching DC motor directions will cause power disruptions and disconnection to the I2C connection between the Raspberry Pi and CRICKIT HAT. The 90% strength provides a high level of throttle and a good level of safety to protect against these problems. You could use even higher values, but the mentioned values have been tested and are reliable in practice.

3.5 *Using environment variables for configuration*

Frequently, some configuration values, specific to a particular machine or hardware device, will need to be set and read by a Python application. Some examples are security credentials or configurable settings that should not have values hard-coded directly into your Python scripts.

In our case, our robot needs a power configuration setting for each of the two DC motors. In the physical world setting, the same throttle on two motors often won't make them move at exactly the same speed. This is due to slight variations in the physical motors. Because one motor will often spin a little faster than the other, it will make the robot veer a little to the left or right instead of moving in a perfectly straight line. The solution is to tweak the throttle settings between the two motors to get them spinning at similar speeds. Then the robot will drive more on a straight path when moving forward. We will create two configuration values to adjust the power of each motor. This is a pragmatic and simple solution that will meet the needs of the projects in this book. A more advanced solution would require us to add hardware sensors and logic in our software to constantly adjust the power on the two motors by taking sensor data into account to keep the robot driving straight.

To solve this problem, we implement a common technique using environment variables in Linux to set our configuration values and then read and use these values in Python. The solution should meet the following requirements:

- Configuration values should be read in Python from a specific set of named environment variables.
- Configuration values should be optional, and if environment variables are not set, they should fall back to specific default values.
- All environment variables are string values. Type casting should be performed to set them to an appropriate data type, such as floating-point values.

Environment variables can be set and viewed in a terminal session. This can be a local terminal session or a remote session over SSH. The default terminal or shell software on the Raspberry Pi OS is called Bash. For more details and help with terminal use and software, see the Raspberry Pi Documentation (https://raspberrypi.com/documentation/usage/terminal).

First, we will define the naming of the environment variables and how they will be set. It is often a good idea to let all your environment variables start with the same prefix. This way, when listing all the existing environment variables, it will be easy to find the ones that are ours. We will use the prefix `ROBO_` for all our variables. Run the command `$ env`.

Execute this command to set a new environment variable that will adjust the amount of power given to the right DC motor. The value of `0.8` will make the throttle lower for the right DC motor and 80% of the normal throttle to slow down the right motor. This could be an adjustment made when you find your right motor moving faster than the left one and want to slow it down so that the two motors have similar speeds:

```
$ export ROBO_DC_ADJUST_R="0.8"
```

When executing the `env` command again, you should see our new variable in the output. We can take the output of this command and use the Bash | feature to pipe the output into another command that will filter the output. The `grep` command filters the output and only shows lines that have the `ROBO_` text in them. We can run the following command to filter the output of the `env` command and only list our variables:

```
$ env | grep ROBO_
```

These values are only available in our current Bash session. If you open a new session or reboot the machine, these values will be lost. To make the environment variable permanent, it should be placed in your .bashrc file. Edit this file and add the `export` line. Now open a new Bash session and confirm the value has been set.

We can now dive into a Python REPL (read–evaluate–print loop) and start reading the values from these environment variables. We will import the `os` module and then start accessing the values:

```
>>> import os
>>> os.environ['ROBO_DC_ADJUST_R']
'0.8'
```

When we access a value that has not been set, a `KeyError` exception will be raised:

```
>>> os.environ['ROBO_DC_ADJUST_L']
Traceback (most recent call last):
  File "<stdin>", line 1, in <module>
  File "/usr/lib/python3.9/os.py", line 679, in __getitem__
    raise KeyError(key) from None
KeyError: 'ROBO_DC_ADJUST_L'
```

The way to deal with optional values is to use the `get` method that provides default values when the environment variable does not exist:

```
>>> os.environ.get('ROBO_DC_ADJUST_L', '1')
'1'
>>> os.environ.get('ROBO_DC_ADJUST_R', '1')
'0.8'
```

We can now typecast our variable into a `float`:

```
>>> float(os.environ.get('ROBO_DC_ADJUST_R', '1'))
0.8
```

Now that we have everything in place, we can upgrade our previous implementation of the `forward` function with these new changes. It is important to note that we don't have to set both environment variables, as each of them is optional. We will save our two configuration values into variables:

```
DC_ADJUST_R = float(os.environ.get('ROBO_DC_ADJUST_R', '1'))
DC_ADJUST_L = float(os.environ.get('ROBO_DC_ADJUST_L', '1'))
```

We will keep our power adjustment values in a dictionary called `ADJUST` so that they can be accessed more easily:

```
ADJUST = dict(R=DC_ADJUST_R, L=DC_ADJUST_L)
```

In a similar fashion, we will access our DC motor objects through a dictionary called `MOTOR`:

```
MOTOR = dict(R=crickit.dc_motor_1, L=crickit.dc_motor_2)
```

The implementation of the `set_throttle` function can now be updated to receive the name of the motor as a string and apply a throttle value that gets adjusted based on the values in `ADJUST`:

```
def set_throttle(name, value):
    MOTOR[name].throttle = value * ADJUST[name]
```

Finally, our `forward` function can be updated to refer to motors using the values `'R'` and `'L'`:

```
def forward():
    set_throttle('R', 0.9)
    set_throttle('L', 0.9)
    time.sleep(0.2)
```

```
    set_throttle('R', 0)
    set_throttle('L', 0)
```

The full script can be saved as `envforward.py` on the Pi and then executed.

Listing 3.2 `envforward.py`: Reading configuration values from environment variables

```
#!/usr/bin/env python3
from adafruit_crickit import crickit
import time
import os

DC_ADJUST_R = float(os.environ.get('ROBO_DC_ADJUST_R', '1'))
DC_ADJUST_L = float(os.environ.get('ROBO_DC_ADJUST_L', '1'))
ADJUST = dict(R=DC_ADJUST_R, L=DC_ADJUST_L)
MOTOR = dict(R=crickit.dc_motor_1, L=crickit.dc_motor_2)

def set_throttle(name, value):
    MOTOR[name].throttle = value * ADJUST[name]

def forward():
    set_throttle('R', 0.9)
    set_throttle('L', 0.9)
    time.sleep(0.2)
    set_throttle('R', 0)
    set_throttle('L', 0)

forward()
```

When the script is run, the robot will move forward based on the specific power adjustments defined for each motor in the environment configuration values. This will enable power adjustments on each wheel and allow it to drive in more of a straight line.

Going deeper: The physics of robotic motion

As your robot projects tackle more complex tasks in challenging environments, the topic of the physics of how robots move becomes increasingly important. For example, if your robot needs to drive on a variety of surfaces that might be slippery at times, then a traction control system can be incorporated into the robot to handle these different surfaces.

Another scenario might be driving the robot over a surface that is not level but has a slope. The slope might be downward or upward. If the slope is upward, we may want to provide more power to the DC motors to achieve the same speed we would have on a level surface. If, on the other hand, the robot is driving downward, then we would want to reduce the power provided to the DC motors so that we don't go too fast. This type of control is a standard feature that is part of the cruise control systems found in many cars. The same systems can be applied to our robots. We would need to add sensors to measure our speed and adjust power accordingly. The following figure provides an illustration of the additional power and force that must be given to the motors when robots drive uphill to counter the force of gravity.

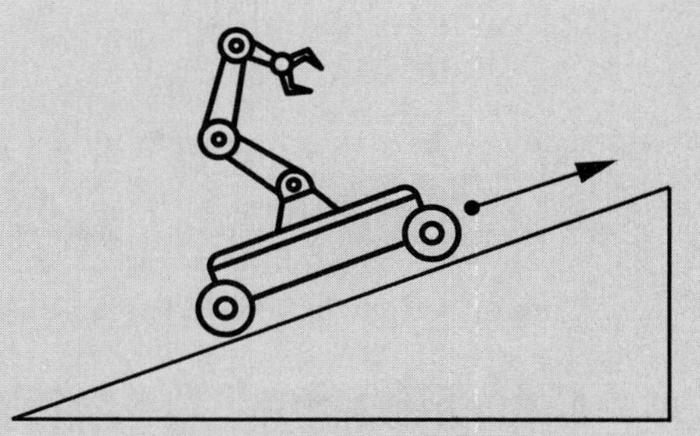

Driving on a slope: when driving up a slope, more power needs to be given to the motors.

The guide on electric traction and steering for robotic vehicles provided by Allied Motion (http://mng.bz/ZRoZ) is an excellent reference on the topics of both steering and traction control in robotics. Topics such as the characteristics of the wheels and operating surfaces are covered. It also includes a comparison of the different ways in which traction solutions can be created with their associated tradeoffs.

3.6 Controlling the speed and duration of movements

In the next upgrade, we will add the ability to control the speed and duration at which we move the robot forward. Currently, the duration is hard-coded in the `forward` function and is set at 0.2 seconds. We will add an optional argument to the function so that it will still default to 0.2 seconds, but the code calling the function can provide other values. The speed at which the robot moves forward can be controlled by changing the level of throttle provided to the motors. We will define three speed settings—low, medium, and high—and then one of these levels can be specified when calling movement functions.

We will add the optional `duration` argument to control how long we run the motors:

```
def forward(duration=0.2):
    set_throttle('R', 0.9)
    set_throttle('L', 0.9)
    time.sleep(duration)
    set_throttle('R', 0)
    set_throttle('L', 0)
```

The `THROTTLE_SPEED` dictionary will map the three speed levels to their associated throttle levels. The speed level 0 is used to stop the motors:

```
THROTTLE_SPEED = {0: 0, 1: 0.5, 2: 0.7, 3: 0.9}
```

We can now update our `forward` function to set the desired speed:

```
def forward(duration=0.2, speed=3):
    set_throttle('R', speed)
    set_throttle('L', speed)
```

```
    time.sleep(duration)
    set_throttle('R', 0)
    set_throttle('L', 0)
```

The `set_throttle` function will now use the new `THROTTLE_SPEED` dictionary too:

```
def set_throttle(name, speed):
    MOTOR[name].throttle = THROTTLE_SPEED[speed] * ADJUST[name]
```

The full script can be saved as `speedforward.py` on the Pi and then executed.

Listing 3.3 `speedforward.py`: Controlling the speed of motors moving forward

```
#!/usr/bin/env python3
from adafruit_crickit import crickit
import time
import os

DC_ADJUST_R = float(os.environ.get('ROBO_DC_ADJUST_R', '1'))
DC_ADJUST_L = float(os.environ.get('ROBO_DC_ADJUST_L', '1'))
ADJUST = dict(R=DC_ADJUST_R, L=DC_ADJUST_L)
MOTOR = dict(R=crickit.dc_motor_1, L=crickit.dc_motor_2)
THROTTLE_SPEED = {0: 0, 1: 0.5, 2: 0.7, 3: 0.9}

def set_throttle(name, speed):
    MOTOR[name].throttle = THROTTLE_SPEED[speed] * ADJUST[name]

def forward(duration=0.2, speed=3):
    set_throttle('R', speed)
    set_throttle('L', speed)
    time.sleep(duration)
    set_throttle('R', 0)
    set_throttle('L', 0)

print('move forward for 0.5 seconds')
forward(duration=0.5)

for speed in [1, 2, 3]:
    print('move forward at speed:', speed)
    forward(speed=speed)
```

The script makes some function calls to `forward` to demonstrate the new functionality. It will move the robot for a custom duration of half a second. Then it will move the robot forward at each of the three speed levels. The robot will stop moving once the last call to the `forward` function is made because the function ends by stopping both motors.

3.7 Moving backward

Now that we have implemented a function for forward movement, we will move on to implement a function to make the robot move backward. These are some of the main movement functions we need to implement to achieve a full range of motion.

First, we will enhance the `set_throttle` function with a `factor` argument. This argument will be used to control whether the throttle will have a positive value to move the motor forward or a negative value to move the motor backward:

```
def set_throttle(name, speed, factor=1):
    MOTOR[name].throttle = THROTTLE_SPEED[speed] * ADJUST[name] * factor
```

Next, we need to implement the new `backward` function. It is very similar to the `forward` function, with the main difference being the value of the `factor` parameter:

```
def backward(duration=0.2, speed=3):
    set_throttle('R', speed, factor=-1)
    set_throttle('L', speed, factor=-1)
    time.sleep(duration)
    set_throttle('R', 0)
    set_throttle('L', 0)
```

The full script can be saved as `backward.py` on the Pi and then executed.

Listing 3.4 `backward.py`: Making the motors move in a backward direction

```
#!/usr/bin/env python3
from adafruit_crickit import crickit
import time
import os

DC_ADJUST_R = float(os.environ.get('ROBO_DC_ADJUST_R', '1'))
DC_ADJUST_L = float(os.environ.get('ROBO_DC_ADJUST_L', '1'))
ADJUST = dict(R=DC_ADJUST_R, L=DC_ADJUST_L)
MOTOR = dict(R=crickit.dc_motor_1, L=crickit.dc_motor_2)
THROTTLE_SPEED = {0: 0, 1: 0.5, 2: 0.7, 3: 0.9}

def set_throttle(name, speed, factor=1):
    MOTOR[name].throttle = THROTTLE_SPEED[speed] * ADJUST[name] * factor

def forward(duration=0.2, speed=3):
    set_throttle('R', speed)
    set_throttle('L', speed)
    time.sleep(duration)
    set_throttle('R', 0)
    set_throttle('L', 0)

def backward(duration=0.2, speed=3):
    set_throttle('R', speed, factor=-1)
    set_throttle('L', speed, factor=-1)
    time.sleep(duration)
    set_throttle('R', 0)
    set_throttle('L', 0)

for i in range(3):
    forward()
    time.sleep(1)
    backward()
    time.sleep(1)
```

The script demonstrates the use of the new function by moving the robot forward and backward three times.

3.8 Turning right

Turning right requires providing different throttle levels to the left and right motor. To better understand the forces at play, we can have a look at the wheel layout in figure 3.4. The left and right wheels have the motors attached to them and can have varying levels of throttle applied. The center wheel is a caster ball that can move freely in any direction. The figure shows that to turn right, we should apply a stronger throttle to the left wheel motor. This will make the left wheel turn faster and thus turn the robot right.

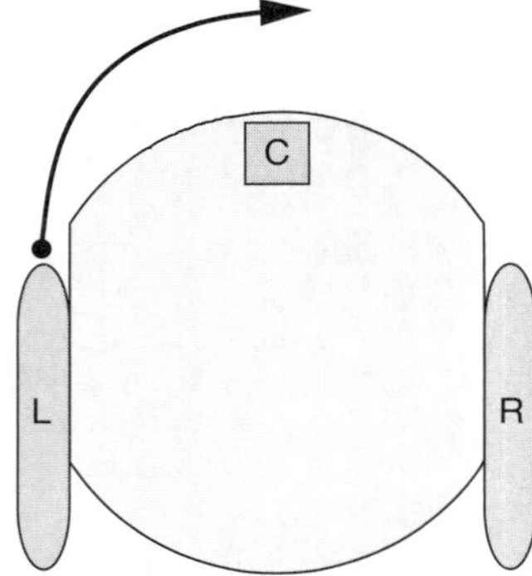

Figure 3.4 Wheel layout: a stronger throttle on the left wheel turns the robot right.

Now we have everything we need to implement the function to move the robot right.

The new `right` function has some similarities to previous functions. Essentially, we are moving in a forward direction but turning right. This is done by giving twice as much throttle to the left wheel:

```
def right(duration=0.2, speed=3):
    set_throttle('R', speed, factor=0.5)
    set_throttle('L', speed)
    time.sleep(duration)
    set_throttle('R', 0)
    set_throttle('L', 0)
```

The full script can be saved as `right.py` on the Pi and then executed.

Listing 3.5 `right.py`: Turning the robot right

```
#!/usr/bin/env python3
from adafruit_crickit import crickit
import time
import os

DC_ADJUST_R = float(os.environ.get('ROBO_DC_ADJUST_R', '1'))
DC_ADJUST_L = float(os.environ.get('ROBO_DC_ADJUST_L', '1'))
ADJUST = dict(R=DC_ADJUST_R, L=DC_ADJUST_L)
MOTOR = dict(R=crickit.dc_motor_1, L=crickit.dc_motor_2)
THROTTLE_SPEED = {0: 0, 1: 0.5, 2: 0.7, 3: 0.9}

def set_throttle(name, speed, factor=1):
    MOTOR[name].throttle = THROTTLE_SPEED[speed] * ADJUST[name] * factor

def forward(duration=0.2, speed=3):
    set_throttle('R', speed)
```

```
    set_throttle('L', speed)
    time.sleep(duration)
    set_throttle('R', 0)
    set_throttle('L', 0)

def backward(duration=0.2, speed=3):
    set_throttle('R', speed, factor=-1)
    set_throttle('L', speed, factor=-1)
    time.sleep(duration)
    set_throttle('R', 0)
    set_throttle('L', 0)

def right(duration=0.2, speed=3):
    set_throttle('R', speed, factor=0.5)
    set_throttle('L', speed)
    time.sleep(duration)
    set_throttle('R', 0)
    set_throttle('L', 0)

right(1)
```

The script calls the `right` function to make the robot turn right for 1 second.

3.9 Moving left and spinning in either direction

We can now implement a complete set of functions to perform all the movements needed for our robot. Here are the requirements we would like to be fulfilled by our set of functions:

- Creating a set of Python functions to make the robot move forward, backward, right, and left, as well as spin right and left.
- Each of these functions should allow us to set the duration and speed of the movement operation.

We have written much of what we need to complete our set of functions. We can now implement the remaining three movement functions. Figure 3.5 shows the direction of throttle that needs to be applied on each motor to make the robot spin to the right.

The new `left` function is essentially just like the `right` function, except with the stronger throttle being put on the right wheel so that the robot will turn left:

```
def left(duration=0.2, speed=3):
    set_throttle('R', speed)
    set_throttle('L', speed, factor=0.5)
    time.sleep(duration)
    set_throttle('R', 0)
    set_throttle('L', 0)
```

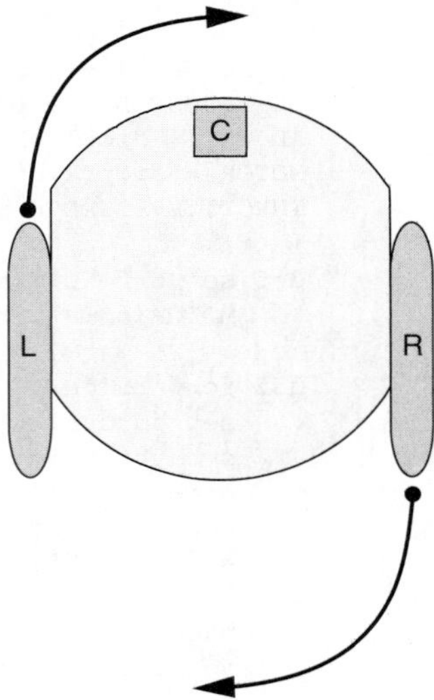

Figure 3.5 Spinning: throttle needs to be applied on the left and right motors to make the robot spin.

The `spin_right` function will make the robot spin in place by having the wheels turn at full speed in opposite directions. The right wheel will spin backward, and the left wheel will spin forward, making the robot spin clockwise:

```
def spin_right(duration=0.2, speed=3):
    set_throttle('R', speed, factor=-1)
    set_throttle('L', speed, factor=1)
    time.sleep(duration)
    set_throttle('R', 0)
    set_throttle('L', 0)
```

The `spin_left` function will spin the robot, but this time in an anticlockwise direction:

```
def spin_left(duration=0.2, speed=3):
    set_throttle('R', speed, factor=1)
    set_throttle('L', speed, factor=-1)
    time.sleep(duration)
    set_throttle('R', 0)
    set_throttle('L', 0)
```

The full script can be saved as `fullmotor.py` on the Pi and then executed.

Listing 3.6 `fullmotor.py`: Creating functions to perform all supported robot movements

```
#!/usr/bin/env python3
from adafruit_crickit import crickit
import time
import os

DC_ADJUST_R = float(os.environ.get('ROBO_DC_ADJUST_R', '1'))
DC_ADJUST_L = float(os.environ.get('ROBO_DC_ADJUST_L', '1'))
ADJUST = dict(R=DC_ADJUST_R, L=DC_ADJUST_L)
MOTOR = dict(R=crickit.dc_motor_1, L=crickit.dc_motor_2)
THROTTLE_SPEED = {0: 0, 1: 0.5, 2: 0.7, 3: 0.9}

def set_throttle(name, speed, factor=1):
    MOTOR[name].throttle = THROTTLE_SPEED[speed] * ADJUST[name] * factor

def forward(duration=0.2, speed=3):
    set_throttle('R', speed)
    set_throttle('L', speed)
    time.sleep(duration)
    set_throttle('R', 0)
    set_throttle('L', 0)

def backward(duration=0.2, speed=3):
    set_throttle('R', speed, factor=-1)
    set_throttle('L', speed, factor=-1)
    time.sleep(duration)
    set_throttle('R', 0)
    set_throttle('L', 0)

def right(duration=0.2, speed=3):
    set_throttle('R', speed, factor=0.5)
```

```
    set_throttle('L', speed)
    time.sleep(duration)
    set_throttle('R', 0)
    set_throttle('L', 0)

def left(duration=0.2, speed=3):
    set_throttle('R', speed)
    set_throttle('L', speed, factor=0.5)
    time.sleep(duration)
    set_throttle('R', 0)
    set_throttle('L', 0)

def spin_right(duration=0.2, speed=3):
    set_throttle('R', speed, factor=-1)
    set_throttle('L', speed, factor=1)
    time.sleep(duration)
    set_throttle('R', 0)
    set_throttle('L', 0)

def spin_left(duration=0.2, speed=3):
    set_throttle('R', speed, factor=1)
    set_throttle('L', speed, factor=-1)
    time.sleep(duration)
    set_throttle('R', 0)
    set_throttle('L', 0)

if __name__ == "__main__":
    left(1)
    spin_right(0.5)
    spin_left(0.5)
```

The script now has all the movement functions implemented. It can even be imported as a Python module and used by other Python scripts. When the script is run directly, it will make the robot turn left and then spin right and left. If, however, the module is imported to be used as a library, then it won't perform those demonstration movements. This is achieved by checking the value of the `__name__` variable to detect whether the Python code is being run directly or imported as a library.

3.10 Refactoring by finding common logic

Code *refactoring* is the process of changing or simplifying how your application is implemented internally without changing any of its external behavior. In our case, we want to simplify the implementation of the `motor` library without changing any of the names of the functions or the arguments they receive. We will do this by making a simpler, more readable, and more maintainable version of this library.

One way to refactor code is to look for logic that is similar or shared between different functions and then centralize that logic in one function to avoid duplication. We can see that our movement functions all share a very similar structure of setting the throttle on the left and right motors, sleeping for some period, and then stopping the throttle on both motors. Let's create a centralized function to implement this functionality and then have the other functions call the main centralized function.

The centralized movement function must take the duration and speed arguments like the other functions, but it also receives the factor values for both the left and right motors, as these vary within different functions. With this implementation, it should meet the needs of all the functions we have implemented:

```
def movement(duration=0.2, speed=3, factor_r=1, factor_l=1):
    set_throttle('R', speed, factor_r)
    set_throttle('L', speed, factor_l)
    time.sleep(duration)
    set_throttle('R', 0)
    set_throttle('L', 0)
```

We can now call the movement function from a new version of the forward function. The implementation is very simple and essentially passes all its arguments as is. We can implement a similar set of changes to migrate all the old functions to use the new approach:

```
def forward(duration=0.2, speed=3):
    movement(duration, speed)
```

The full script can be saved as commonmotor.py on the Pi and then imported.

Listing 3.7 commonmotor.py: Consolidating common logic into a common function

```
#!/usr/bin/env python3
from adafruit_crickit import crickit
import time
import os

DC_ADJUST_R = float(os.environ.get('ROBO_DC_ADJUST_R', '1'))
DC_ADJUST_L = float(os.environ.get('ROBO_DC_ADJUST_L', '1'))
ADJUST = dict(R=DC_ADJUST_R, L=DC_ADJUST_L)
MOTOR = dict(R=crickit.dc_motor_1, L=crickit.dc_motor_2)
THROTTLE_SPEED = {0: 0, 1: 0.5, 2: 0.7, 3: 0.9}

def set_throttle(name, speed, factor=1):
    MOTOR[name].throttle = THROTTLE_SPEED[speed] * ADJUST[name] * factor

def movement(duration=0.2, speed=3, factor_r=1, factor_l=1):
    set_throttle('R', speed, factor_r)
    set_throttle('L', speed, factor_l)
    time.sleep(duration)
    set_throttle('R', 0)
    set_throttle('L', 0)

def forward(duration=0.2, speed=3):
    movement(duration, speed)

def backward(duration=0.2, speed=3):
    movement(duration, speed, factor_r=-1, factor_l=-1)

def right(duration=0.2, speed=3):
    movement(duration, speed, factor_r=0.5)
```

```
def left(duration=0.2, speed=3):
    movement(duration, speed, factor_l=0.5)

def spin_right(duration=0.2, speed=3):
    movement(duration, speed, factor_r=-1, factor_l=1)

def spin_left(duration=0.2, speed=3):
    movement(duration, speed, factor_r=1, factor_l=-1)
```

The implementation for each function is much simpler now. This will make the code much more maintainable, as all the real work is being done by the `set_throttle` and `movement` functions. It is also more readable, as each call is essentially just changing some parameters on the `movement` function call.

> **Going deeper: Code refactoring**
>
> The process of code refactoring is an important part of software development. The Agile Alliance (https://www.agilealliance.org/glossary/refactoring) has a nice definition of what refactoring is and some of its benefits. When we first implement a piece of software, our goal is often just to get the thing working. Once we have it working, the code base will naturally grow as we add more features to the software. Refactoring is when we make the time to step back and not add any new functionality but think of ways of cleaning up or simplifying our code base.
>
> One of the benefits of a cleaner code base is improved maintainability. In the long run, having a cleaner code base will save us a lot of time as the code becomes more manageable. It can make the software more reliable by making it easier to find and fix bugs.
>
> Another important concept related to refactoring is the design principle of Don't Repeat Yourself (DRY). When we apply the DRY principle to our code, we want to avoid repetition in our code and logic. In section 3.10, we found logic that was duplicated and reduced that duplication by integrating it into one common function. Many good-quality frameworks and libraries apply a DRY philosophy to their software design and support creating projects that avoid repetition. The Python Django web framework is a good example of this, and its documentation (https://docs.djangoproject.com) has a page dedicated to its design philosophies. It mentions DRY and other design principles that create a cleaner code base.

3.11 Refactoring by using functools

The `functools` module is part of the Python standard library, and it provides several different functionalities around callable objects and functions. Specifically, we will use `partial` to simplify the way we define functions, which is a perfect tool for scenarios where one function is essentially very similar to a call to another function, as is our case.

The way `partial` works is that it takes some existing function as its first argument and a set of positional and keyword arguments. Then it returns a new function that would call the original function with those provided arguments. We can use it to simplify our function definitions. For more details on `partial` and example functions created using `partial`, check the Python documentation (https://docs.python.org/3/library/functools.html) on the `functools` module.

We first import `partial` from the `functools` module:

```
from functools import partial
```

The new definition of the `forward` function is now essentially a direct call to `movement`, as it has default values that directly map to the defaults of the `forward` function:

```
forward = partial(movement)
```

In the case of `backward`, the only change to the arguments of `movement` is to set both the left and right motors to turn in reverse:

```
backward = partial(movement, factor_r=-1, factor_l=-1)
```

We continue the process for `right` and `left`, which use all the default values except for reducing the right and left motor speeds to make the robot turn:

```
right = partial(movement, factor_r=0.5)
left = partial(movement, factor_l=0.5)
```

The `spin_right` and `spin_left` are created using a similar approach:

```
spin_right = partial(movement, factor_r=-1, factor_l=1)
spin_left = partial(movement, factor_r=1, factor_l=-1)
```

We also add a `noop` function that will help us with performance testing in later chapters:

```
noop = lambda: None
```

The full script can be saved as `motor.py` on the Pi and then imported.

Listing 3.8 `motor.py`: Simplifying the way functions are defined

```
#!/usr/bin/env python3
from adafruit_crickit import crickit
import time
import os
from functools import partial

DC_ADJUST_R = float(os.environ.get('ROBO_DC_ADJUST_R', '1'))
DC_ADJUST_L = float(os.environ.get('ROBO_DC_ADJUST_L', '1'))
ADJUST = dict(R=DC_ADJUST_R, L=DC_ADJUST_L)
MOTOR = dict(R=crickit.dc_motor_1, L=crickit.dc_motor_2)
THROTTLE_SPEED = {0: 0, 1: 0.5, 2: 0.7, 3: 0.9}

def set_throttle(name, speed, factor=1):
    MOTOR[name].throttle = THROTTLE_SPEED[speed] * ADJUST[name] * factor

def movement(duration=0.2, speed=3, factor_r=1, factor_l=1):
    set_throttle('R', speed, factor_r)
    set_throttle('L', speed, factor_l)
```

```
    time.sleep(duration)
    set_throttle('R', 0)
    set_throttle('L', 0)

forward = partial(movement)
backward = partial(movement, factor_r=-1, factor_l=-1)
right = partial(movement, factor_r=0.5)
left = partial(movement, factor_l=0.5)
spin_right = partial(movement, factor_r=-1, factor_l=1)
spin_left = partial(movement, factor_r=1, factor_l=-1)
noop = lambda: None
```

We can take the final version of our `motor` library for a spin by starting a Python REPL session in the same path. In the following session, the robot is moved forward and backward. Then it turns right for half a second and then left at the lowest speed. Next, it spins right for a second and spins left for 2 seconds. The last call will perform no operation but should execute successfully with no errors:

```
>>> import motor
>>> motor.forward()
>>> motor.backward()
>>> motor.right(0.5)
>>> motor.left(speed=1)
>>> motor.spin_right(1)
>>> motor.spin_left(2)
>>> motor.noop()
```

You can import this library into all sorts of applications and call different movement functions to create many different projects. You could write a program that responds to user input to move the robot in different directions. Or you could program the robot to drive around an obstacle course with the correct forward and turn movements so as not to collide with any objects.

Going deeper: The power of libraries

It is one thing to write a standalone program and a very different thing to write a library that provides functionality for many different pieces of software. The earlier scripts in this chapter were standalone scripts that could be executed to move the robot around. They were limited in the actions they could perform but were a great starting point. As we built up the functionality, we were able to package it all into a library. As we create the different projects in the book, we will often import the `motor` library from this chapter and use its functionality.

Well-designed libraries can provide a bundle of reusable functionality that we can easily incorporate into our own programs. We can combine them to create all sorts of powerful and complex applications by building on the great works of others or, as Isaac Newton said, "by standing on the shoulders of giants." In this book, we will use libraries we create as well as a rich set of open source Python libraries that we can easily install and incorporate into our robot projects.

Summary

- Robot chassis kits are a flexible and inexpensive way to build mobile robots.
- The right and left wheels each have their own dedicated DC motor.
- Environment variables in Linux can be used to pass configuration values to Python scripts, which can then be used to configure motor power adjustments.
- Configuration values can be made optional by detecting missing environment variables and setting default values to fall back on.
- By simultaneously turning on the left and right DC motor, forward motion of the robot can be achieved.
- The robot's speed can be controlled by changing the throttle level of DC motors.
- Backward motion can be achieved by reversing the throttle direction of both DC motors.
- To turn the robot right, we need to apply a stronger throttle to the left wheel motor in comparison to the right wheel motor.
- To make the robot spin, we need to apply a throttle on both motors in an opposite direction.
- Refactoring code can simplify how your application is implemented without altering the way the functions in your library get called.
- Creating functions can become faster and simpler using `partial`.

4 Creating a robot shell

This chapter covers

- The basics of creating interactive custom shells in Python
- Creating a command loop for moving the robot forward and backward
- Handling command arguments in shell commands
- Centralizing argument-handling logic in code shells
- Executing custom shell scripts in Python

This chapter will teach you how to create a custom interactive REPL (read–evaluate–print loop) robot shell. Shells provide a powerful interactive interface that enables direct interaction with software or, in this case, robotic hardware. They are a tried-and-true method for user interaction, and the Python standard library provides a built-in functionality to create custom shells. The chapter starts with a simple robotic shell and then progresses to add more movement functions with more customized options. The chapter ends by showing how a set of commands can be saved and run in one go in the shell, as is done in many other shells.

4.1 What's a REPL or shell?

A REPL or command-line shell is a program that loops endlessly, waiting to receive user input, which is then taken and executed, and the output is printed as needed. They are also called *line-oriented command interpreters* because they take commands as a line of user input and interpret or execute the given line. Figure 4.1 illustrates the three states that a REPL goes through.

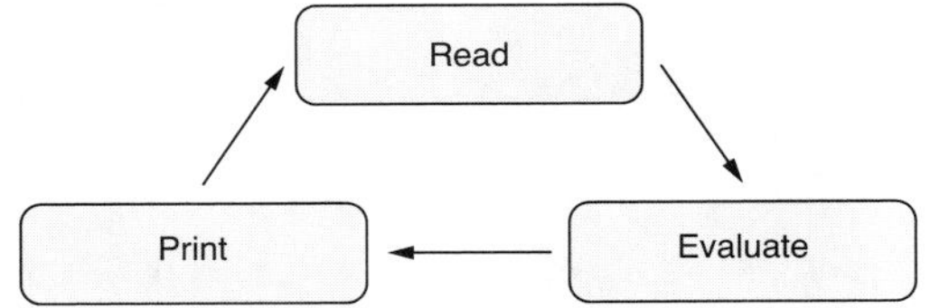

Figure 4.1 Read–evaluate–print loop: the REPL goes endlessly between the read–evaluate–print states.

4.2 Benefits of a REPL

In the book so far, we have already interacted with two very popular programs that provide a REPL interface: Python and Bash. Here are some of the features and benefits of the shells created in this chapter:

- Fast and direct execution of commands.
- Simple line-by-line execution.
- The live help command lists available commands.
- Tab auto-completion of commands.
- Command history, which can be accessed with up and down keys.
- Optional arguments for each command.
- Execution of commands in a script file.

These features will be used by the robot shell created in this chapter. With our robot shell, we will be able to issue movement commands quickly and easily to our robot from the terminal. We can also use the command history feature to replay past movements we have applied to the robot.

4.3 Hardware stack

Figure 4.2 shows the hardware stack discussed, with the specific components used in this chapter highlighted. The REPL utilizes keyboard interaction by employing Tab auto-completion and arrow keys to access the command history.

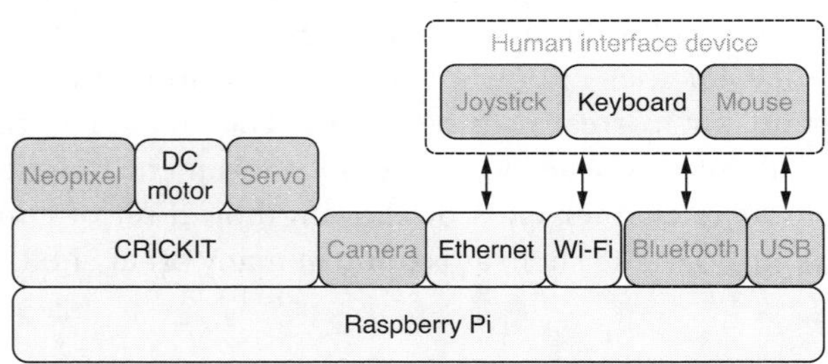

Figure 4.2 Hardware stack: the robot shell controls the DC motor through the CRICKIT board.

The robot shell will be running on the Raspberry Pi hardware. The commands executed through the shell will communicate with the CRICKIT board, which will then send the signals to the DC motors to make the requested motor movements.

4.4 Software stack

Details of the specific software used in this chapter are illustrated in figure 4.3 and described in the text that follows. With each new application, we will add more features and enhancements to our shell implementations.

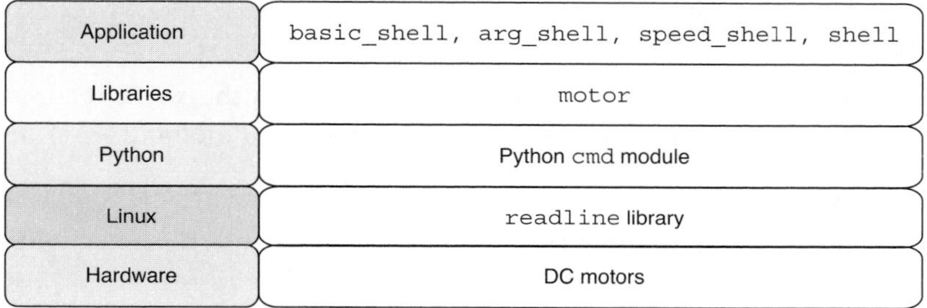

Figure 4.3 Software stack: the robot shell will run on the Python interpreter.

Python and Bash both have a REPL feature, and both run directly on Linux using the `readline` library. The `motor` module from the previous chapter will be used to control the DC motors. The robot shell in this chapter will be implemented using Python, so it will be running on top of Python. Users do not have to worry about how it is implemented and can interact with it as with any other shell program on their computer.

4.5 Creating the robot shell

In this section, we will use Python to write a program that implements a custom REPL robot shell to obtain basic robot actions of forward and backward movements. This script requires two modules, which we import as shown in the following code. The `cmd` module is part of the Python standard library and provides a framework for creating REPL applications such as the robot shell. The `cmd` module (https://docs.python.org/3/library/cmd.html) documentation is an excellent resource for learning more about the library. The `motor` module is the library created in the previous chapter for controlling the robot:

```
import cmd
import motor
```

Next, we define our `RobotShell` class, which is a subclass of `cmd.Cmd`. The `intro` provides a welcome message when the shell is first started. The `prompt` text appears when the user is prompted, indicating they are now in the robot shell. Each method that starts with the name `do_` is automatically called to handle its related command. In this way, the `do_forward` method gets called to handle `forward` commands. Each time it is called, it moves the robot forward by calling `motor.forward`:

```
class RobotShell(cmd.Cmd):
    intro = 'Welcome to the robot shell. Type help or ? to list commands.'
    prompt = '(robot) '

    def do_forward(self, line):
        motor.forward()
```

Next, we define the `do_backward` method to handle the `backward` command:

```
    def do_backward(self, line):
        motor.backward()
```

The final line of code will run the event loop that starts the shell, read commands from user input, and execute the corresponding command method:

```
RobotShell().cmdloop()
```

The full script can be saved as `basic_shell.py` on the Pi and then executed.

Listing 4.1 `basic_shell.py`: Providing basic robot movements in a shell

```
#!/usr/bin/env python3
import cmd
import motor

class RobotShell(cmd.Cmd):
    intro = 'Welcome to the robot shell. Type help or ? to list commands.'
    prompt = '(robot) '

    def do_forward(self, line):
        motor.forward()

    def do_backward(self, line):
        motor.backward()

RobotShell().cmdloop()
```

When executing the script, make sure that `motor.py` is in the same directory so that it can be imported by `basic_shell.py`. The following code shows an example of a session in the robot shell where `help` and movement commands have been executed:

```
$ basic_shell.py
Welcome to the robot shell. Type help or ? to list commands.
(robot) help

Documented commands (type help <topic>):
========================================
help

Undocumented commands:
======================
backward  forward
```

```
(robot) forward
(robot) backward
(robot) backward
(robot) forward
```

We can run the shell remotely over an SSH connection like any other Python script. When running the shell, the Tab key can be pressed to use the auto-complete feature, and the up and down arrows can be used to access the command history feature. Press the F key and then the Tab key to have the command `forward` get auto-completed. When you finish using the robot shell, you can press CTRL+C to exit like you would in other programs.

Going deeper: Importing libraries

In this chapter, we are building on the code from the previous chapter by importing the `motor` library. We can keep all the scripts and libraries we create in the `/home/robo/bin/` directory to simplify the process of importing modules. But where are the other libraries we have been using located, and how does the Python interpreter figure out where to find them when we import modules?

We can pop into the REPL to get answers to these questions. We import the `sys` module and then inspect the contents of `sys.path`:

```
>>> import sys
>>> sys.path
['', '/usr/lib/python39.zip', '/usr/lib/python3.9',
 '/usr/lib/python3.9/lib-dynload',
 '/home/robo/pyenv/lib/python3.9/site-packages']
```

In `sys.path`, there is a list of strings that are the paths to be searched when importing modules. If we check out these directories, we will find the modules that we have been importing in the book. For example, the location of the `cmd` module that is part of the standard library can be listed using

```
$ ls /usr/lib/python3.9/cmd.py
/usr/lib/python3.9/cmd.py
```

We can open this file and investigate its source code like any other Python script. We can also find the location of third-party libraries that we have installed in our virtual environment using the `pip` command. Here is the location of the Adafruit CRICKIT library that we used to control the DC motors:

```
$ ls /home/robo/pyenv/lib/python3.9/site-packages/adafruit_crickit.py
/home/robo/pyenv/lib/python3.9/site-packages/adafruit_crickit.py
```

We can see that the libraries in the Python standard library are in the system location shared by all virtual environments, while the third-party packages installed in our virtual environment are all located in the `/home/robo/pyenv` location we created for our virtual environment. For further details on importing packages in Python, the documentation for the `importlib` (https://docs.python.org/3/library/importlib.html) module is a great resource.

4.6 Handling command arguments

We have implemented a basic robot shell with the two movement commands, `forward` and `backward`. However, they can't handle any arguments provided after the movement command. We will add support for providing the `duration` argument for each movement command. We will also improve the way we exit the shell by properly handling end-of-file (EOF) on input.

We now enhance the `do_forward` method to check if a duration has been provided. The text after the `forward` command will be provided in the `line` argument, which we can parse to get the duration. If a duration is found, it will be converted into a `float` value and used when calling the `motor.forward` function:

```
def do_forward(self, line):
    if line:
        duration = float(line)
        motor.forward(duration)
    else:
        motor.forward()
```

The same process is then applied to the `do_backward` method:

```
def do_backward(self, line):
    if line:
        duration = float(line)
        motor.backward(duration)
    else:
        motor.backward()
```

The `do_EOF` method is added to the class to handle when an EOF condition is encountered in the input data. The method returns a `True` value to signal to the event loop that the shell is to be exited:

```
def do_EOF(self, line):
    return True
```

The full script can be saved as `arg_shell.py` on the Pi and then executed.

Listing 4.2 `arg_shell.py`: Supporting command arguments in the shell

```
#!/usr/bin/env python3
import cmd
import motor

class RobotShell(cmd.Cmd):
    intro = 'Welcome to the robot shell. Type help or ? to list commands.'
    prompt = '(robot) '

    def do_EOF(self, line):
        return True

    def do_forward(self, line):
        if line:
```

```
            duration = float(line)
            motor.forward(duration)
        else:
            motor.forward()

    def do_backward(self, line):
        if line:
            duration = float(line)
            motor.backward(duration)
        else:
            motor.backward()

RobotShell().cmdloop()
```

What follows is an example of a session in the robot shell where movement commands with different durations are called:

```
$ arg_shell.py
Welcome to the robot shell. Type help or ? to list commands.
(robot) forward 0.2
(robot) forward 1
(robot) backward 0.5
(robot) backward
(robot) forward
```

In the example session, the movement commands were called with durations expressed as integers and floats. The commands can also be called without providing any duration that will use the default duration value. When you want to exit the robot shell, you press CTRL+D instead of CTRL+C. This will exit the shell in a much cleaner fashion, as CTRL+D will send an EOF, whereas pressing CTRL+C will end the shell spitting out a `Traceback` error. Using CTRL+D to exit a shell is a standard procedure, and the same process will work with Bash and the Python REPL.

4.7 Adding a speed argument

To support multiple optional arguments, we will have to do a little extra work. As we need to upgrade our argument-handling behavior, it would be better if we don't have to change it in multiple places. So the first step will be to centralize our argument-handing code for all movement commands in one function and then upgrade that function.

We create a new function, `get_kwargs`, that will take the `line` value and return a `dict` object with all the required key-value pairs. The following definition will cover the existing behavior of taking the first optional argument as the value for `duration`:

```
def get_kwargs(line):
    if line:
        return dict(duration=float(line))
    else:
        return dict()
```

We then update the definitions of `do_forward` and `do_backward` to use `get_kwargs`. They call `get_kwargs` and directly use what is returned as the keyword arguments for the function call to their associated movement functions:

```
    def do_forward(self, line):
        motor.forward(**get_kwargs(line))

    def do_backward(self, line):
        motor.backward(**get_kwargs(line))
```

At this stage, we could run the shell, and it would work with the previous behavior. We can now upgrade the `get_kwargs` function and add the handling of the second keyword argument `speed`. This argument is expected to be of type `int`, so it is typecast to that data type:

```
def get_kwargs(line):
    kwargs = dict()
    items = line.split()
    if len(items) > 0:
        kwargs['duration'] = float(items[0])
    if len(items) > 1:
        kwargs['speed'] = int(items[1])
    return kwargs
```

The full script can be saved as `speed_shell.py` on the Pi and then executed.

Listing 4.3 `speed_shell.py`: Controlling movement speed in the shell

```
#!/usr/bin/env python3
import cmd
import motor

def get_kwargs(line):
    kwargs = dict()
    items = line.split()
    if len(items) > 0:
        kwargs['duration'] = float(items[0])
    if len(items) > 1:
        kwargs['speed'] = int(items[1])
    return kwargs

class RobotShell(cmd.Cmd):
    intro = 'Welcome to the robot shell. Type help or ? to list commands.'
    prompt = '(robot) '

    def do_EOF(self, line):
        return True

    def do_forward(self, line):
        motor.forward(**get_kwargs(line))

    def do_backward(self, line):
        motor.backward(**get_kwargs(line))

RobotShell().cmdloop()
```

Here is an example of a session in the upgraded robot shell:

```
$ speed_shell.py
Welcome to the robot shell. Type help or ? to list commands.
(robot) forward 0.2 1
(robot) forward 0.2 3
(robot) backward 0.5 1
(robot) backward 1 2
(robot) backward 0.5
(robot) forward
```

In the example session, the movement commands can now be called with the duration specified and default speed, with a specific duration and speed or with the default duration and speed settings.

4.8 Running robot shell scripts

In this section, the challenge is to make the robot shell have the capability of executing a script of commands and adding the remaining movement commands. We add the methods `do_right`, `do_left`, `do_spin_right`, and `do_spin_left`. They follow the same style as our previous movement methods:

```
    def do_right(self, line):
        motor.right(**get_kwargs(line))

    def do_left(self, line):
        motor.left(**get_kwargs(line))

    def do_spin_right(self, line):
        motor.spin_right(**get_kwargs(line))

    def do_spin_left(self, line):
        motor.spin_left(**get_kwargs(line))
```

When we have the shell execute the commands in a script file, it would be great to get some visual feedback regarding which command is being executed. We can do this by adding a `precmd` method. This method gets called before executing any command. It is a feature that comes as part of the `cmd.Cmd` object. We will use it to print the command that is about to be executed. It must return the value of `line` to have the event loop process the command:

```
    def precmd(self, line):
        print('executing', repr(line))
        return line
```

The full script can be saved as `shell.py` on the Pi and then executed.

Listing 4.4 `shell.py`: Creating a shell supporting all robot movements

```
#!/usr/bin/env python3
import cmd
import motor
```

```
def get_kwargs(line):
    kwargs = dict()
    items = line.split()
    if len(items) > 0:
        kwargs['duration'] = float(items[0])
    if len(items) > 1:
        kwargs['speed'] = int(items[1])
    return kwargs

class RobotShell(cmd.Cmd):
    intro = 'Welcome to the robot shell. Type help or ? to list commands.'
    prompt = '(robot) '

    def do_EOF(self, line):
        return True

    def precmd(self, line):
        print('executing', repr(line))
        return line

    def do_forward(self, line):
        motor.forward(**get_kwargs(line))

    def do_backward(self, line):
        motor.backward(**get_kwargs(line))

    def do_right(self, line):
        motor.right(**get_kwargs(line))

    def do_left(self, line):
        motor.left(**get_kwargs(line))

    def do_spin_right(self, line):
        motor.spin_right(**get_kwargs(line))

    def do_spin_left(self, line):
        motor.spin_left(**get_kwargs(line))

RobotShell().cmdloop()
```

The following text file of commands should be saved as `move.txt` on the Pi:

```
spin_right
spin_left
right
left
forward 0.2 1
forward 0.2 3
backward 0.5 1
backward 0.5
```

As an initial test, we can use `echo` to feed a single command into the robot shell:

```
$ echo forward | shell.py
Welcome to the robot shell. Type help or ? to list commands.
```

```
(robot) executing 'forward'
(robot) executing 'EOF'
```

We can also use `cat` to feed a whole set of saved commands into the robot shell:

```
$ cat move.txt | shell.py
Welcome to the robot shell. Type help or ? to list commands.
(robot) executing 'spin_right'
(robot) executing 'spin_left'
(robot) executing 'right'
(robot) executing 'left'
(robot) executing 'forward 0.2 1'
(robot) executing 'forward 0.2 3'
(robot) executing 'backward 0.5 1'
(robot) executing 'backward 0.5'
(robot) executing 'EOF'
```

In this way, we can design our own shell using a set of commands that best suits our needs. A set of robot shell commands can either be run interactively or saved in a single file to be executed directly by the robot shell.

Going deeper: Enhancing the shell

We can take our shell further and add some more powerful features. In the documentation for `cmd` (https://docs.python.org/3/library/cmd.html), there is one feature that is very useful for our robot use case. The documentation shows how a REPL session can be recorded and then played back later by creating commands that record and play back the session. Let's say we are using the robot shell to move the robot around a physical course with a certain set of movements that work. Instead of having to retype them all, we can record and play back the robot's movements whenever we want.

Another common and powerful feature is to execute robot commands through command line arguments. The Python Module of the Week website is a great resource for diving deeper into different parts of the Python standard modules, and their documentation on the `cmd` module (https://pymotw.com/3/cmd/) shows many different ways of using the module, including how to parse commands from the command line arguments. The Python interpreter itself has this functionality. We have used the REPL before in the book, but we can also evaluate Python code by passing the code as a command line argument to the interpreter directly. What follows is a simple example demonstrating how to use this feature:

```
$ python -c 'print(1+1)'
2
```

A third useful feature we could add to the shell is the ability to retain our command history between shell sessions. Currently, when we use our shell, we can use the up and down arrows to go back through commands we have issued. But once we exit our shell, we lose this history. Other shells such as the Python REPL retain the history between REPL sessions. This is done by saving a history file when we exit the shell and loading it back when we start a new shell. We can see this in action with the Python REPL. Open a REPL and evaluate some Python expressions. Now exit the REPL

(continued)

and open a new one. If you press the up arrow, you will find the commands in your history. We can find the file that stores this history and output its contents using the following commands:

```
$ ls ~/.python_history
/home/robo/.python_history
$ cat ~/.python_history
```

To implement this feature in our robot shell, we would use the `readline` module (https://docs.python.org/3/library/readline.html), which is what is handling the command history functionality of our shell. It has a set of functions that will let us save and load the history to a history file. The Python Module of the Week page for the `readline` module (https://pymotw.com/3/readline/) has an excellent example that implements this feature. We would only need to add a few lines of code to our shell startup to load the history file and then some code when we exit the shell to save the history file.

Summary

- A REPL is a program that loops endlessly, waiting to receive user input.
- The commands executed through the robot shell will communicate with the CRICKIT board, which will then send the signals to the DC motors to make the requested motor movements.
- Python and Bash both have a REPL feature, and both run directly on Linux.
- The `cmd` module is a part of the Python standard library and provides a framework for creating REPL applications such as the robot shell.
- The `do_EOF` method is used to handle EOF conditions encountered in the input data.
- Optional arguments make it possible to call movement commands with duration specified and default speed, with a specific duration and speed, or with default duration and speed settings.
- The `cat` command can be used to feed a set of saved commands into the robot shell.

5 Controlling robots remotely

This chapter covers

- Executing robot commands over the network using SSH
- Creating web services to control robots
- Calling robot web services from Python
- Creating Python-based remote execution functions
- Measuring the execution time of local and remote HTTP commands
- Building a high-performance Python client with low-latency calls

This chapter will teach you how to share your robot over your network so that remote Python clients can issue movement commands to control it. Both SSH and HTTP protocols will be used, meaning that two solutions from a protocol perspective will be offered, each with its own set of benefits and tradeoffs. For the HTTP solution, a simple Python client will be created first, followed by a more complex, higher-performance low-latency client. Furthermore, different techniques for measuring the execution time will be covered as well. This will provide a quantitative basis to compare the performance of different protocols and clients.

Controlling robots remotely is an essential part of many projects, such as using apps on mobile phones and laptops to control robots, as well as controlling a fleet of robots using a central robot server. The projects in this chapter make it possible to control robots in the same room or many miles away. Unlike short-range protocols such as Bluetooth, both SSH and HTTP support short- and long-range connectivity.

5.1 Hardware stack

Figure 5.1 shows the hardware stack, with the specific components used in this chapter highlighted. The robot will be connected to wired networks using the Ethernet port and wireless networks using the Wi-Fi hardware. The Wi-Fi connectivity gives the robot the greatest freedom of movement, allowing it to move around without any attached wires. However, wired Ethernet connections at times can give better performance. This chapter will show how to take network performance measurements so that these two options can be compared.

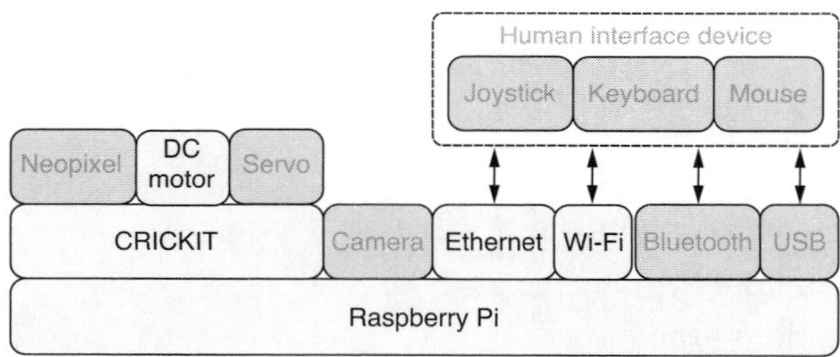

Figure 5.1 Hardware stack: remote clients will connect using either the Ethernet or Wi-Fi hardware.

5.2 Software stack

Details of the specific software used in this chapter are illustrated in figure 5.2 and described in the text that follows. The three main applications in this chapter will

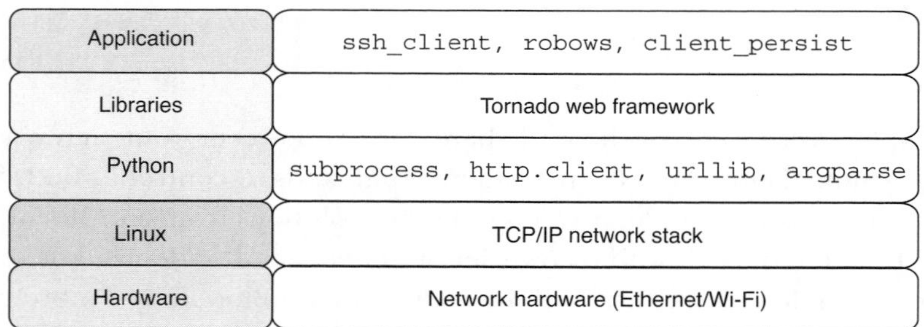

Figure 5.2 Software stack: the Tornado web framework will expose robot commands over the HTTP protocol.

implement a remote client over the SSH protocol (`ssh_client.py`), serving robot web services (`robows.py`), and a web service client using persistent connections (client_persist.py). The Tornado web framework will be used to create the HTTP solution. The `subprocess` and `argparse` Python modules will be used in constructing the SSH solution. The first HTTP client will use the `urllib` module, and then a more advanced version will use the `http.client` module.

Going deeper: Web frameworks

When creating a web application in Python, it is almost always a good idea to use a web framework. There are so many details that need to be taken care of when creating a web application, and web frameworks do a great job of addressing these needs. In Python, we have many great options to choose from, such as the Django (https://www.djangoproject.com) and Flask (https://flask.palletsprojects.com) web frameworks.

In our case, we will use the Tornado web framework (https://www.tornadoweb.org) because it has a special feature that makes it perfect for our needs. Most web frameworks, such as Django and Flask, do not come with a production-ready web application server that can safely control hardware such as our robot motors. Tornado, however, offers such options. It lets our web application run in a single long-running process for the whole life cycle of the web server. This process also gets exclusive access to our robot motors, as it will only allow one web request to move the robot motors at a time. In this way, we can avoid race conditions and keep our web application safe and simple to implement.

5.3 Robot testing tips

For details on assembling and configuring the robot hardware, check the robot assembly guide in appendix C. There are two tips that can help you when working with the assembled robot in this chapter.

The first tip is to place the robot on a stand when initially testing your code base. Figure 5.3 shows the robot placed on a stand that lets its wheels move freely without the robot moving around. In this way, you can safely place the robot on a table during testing and not worry about accidentally driving it off the table and damaging it. This is particularly useful when you have new, untested code that might enter into a state of starting the motors and not

Figure 5.3 Robot stand: for safety, the robot can be placed on a stand.

stopping them, which may send the robot in some direction only to crash into a wall or some other object.

The second tip is to use SlimRun Ethernet cables instead of standard ones. These cables are lighter and thinner than standard network cables, which gives the robot more maneuverability when driving with a wired network connection. Figure 5.4 shows a SlimRun network cable connected to the robot.

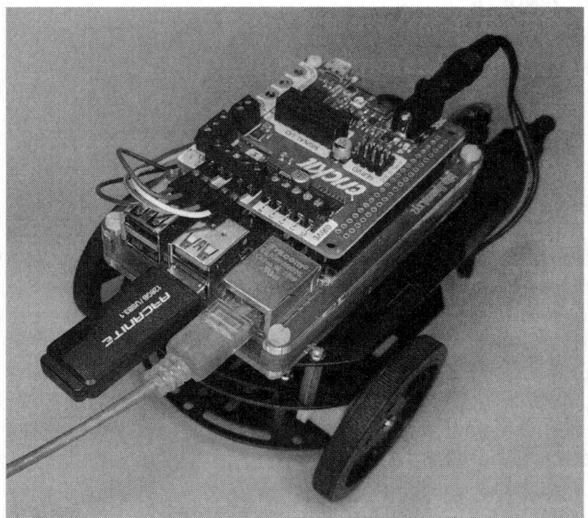

Figure 5.4 Robot network cable: the network cable is connected to the Ethernet port on the robot.

Using these two tips can protect your robot against unnecessary damage and make it more maneuverable over wired connections. Keep your robot safe because a broken robot is no fun.

5.4 Controlling robots over SSH

We will use the SSH protocol for our first solution to control robots over a network. SSH is easier to get started with because we already have our SSH server setup in place, and we will use SSH clients and connections throughout the book to connect to our robot. We need to create some Python code on top of SSH to meet the following requirements:

- Create a Python script that will be executed on the SSH server receiving the movement action and optional movement arguments, and perform the robot movements.
- Create a Python function for the SSH client that receives the name of the movement action and connects to the robot and remotely executes the same movement.

5.4.1 Creating the server-side script

The first step is to import all the necessary modules. We import `ArgumentParser` from the `argparse` module that is part of the Python standard library and that will perform all the heavy lifting for parsing command line arguments. The `motor` module is the library created in chapter 3 for controlling the robot:

```
from argparse import ArgumentParser
import motor
```

The function `parse_args` is then defined, and it takes care of all the command line argument parsing. It first creates an `ArgumentParser` object and then configures the parser. One required argument called `name` will capture the name of the movement. Then the optional `--duration` and `--speed` arguments are configured. They are both configured with their correct data type and a help message. The last line of the function will perform the actual parsing, and it uses the `vars` function to return the result as a `dict` object:

```
def parse_args():
    parser = ArgumentParser(description='robot cli')
    parser.add_argument('name', help='name of movement')
    parser.add_argument('--duration', type=float, help='movement duration')
    parser.add_argument('--speed', type=int, help='movement speed')
    return vars(parser.parse_args())
```

The `main` function will first call `parse_args` and save the result in the `args` variable. The name of the movement function to be called is then removed from `args` and saved in the `name` variable. Now the movement function can be looked up from the `motor` module using `getattr`. The next step is to collect all the optional arguments that have been specified and save them in a dictionary called `kwargs`. Finally, the function to be called is printed and called:

```
def main():
    args = parse_args()
    name = args.pop('name')
    func = getattr(motor, name)
    kwargs = {k: v for k, v in args.items() if v}
    print(f'calling {name} with kwargs {kwargs}')
    func(**kwargs)
```

The full script can be saved as `cli.py` on the Pi and then executed.

Listing 5.1 `cli.py`: Creating command line interface to perform robot movements

```
#!/usr/bin/env python3
from argparse import ArgumentParser
import motor

def parse_args():
    parser = ArgumentParser(description='robot cli')
    parser.add_argument('name', help='name of movement')
```

```
    parser.add_argument('--duration', type=float, help='movement duration')
    parser.add_argument('--speed', type=int, help='movement speed')
    return vars(parser.parse_args())

def main():
    args = parse_args()
    name = args.pop('name')
    func = getattr(motor, name)
    kwargs = {k: v for k, v in args.items() if v}
    print(f'calling {name} with kwargs {kwargs}')
    func(**kwargs)

main()
```

The following code is a session in the terminal demonstrating different calls to the script:

```
$ cli.py
usage: cli.py [-h] [--duration DURATION] [--speed SPEED] name
cli.py: error: the following arguments are required: name
$ cli.py --help
usage: cli.py [-h] [--duration DURATION] [--speed SPEED] name

robot cli

positional arguments:
  name                  name of movement

optional arguments:
  -h, --help            show this help message and exit
  --duration DURATION   movement duration
  --speed SPEED         movement speed
$ cli.py forward
calling forward with kwargs {}
$ cli.py forward --duration=0.5
calling forward with kwargs {'duration': 0.5}
$ cli.py forward --speed=1
calling forward with kwargs {'speed': 1}
$ cli.py forward --duration=0.5 --speed=1
calling forward with kwargs {'duration': 0.5, 'speed': 1}
```

The script is first called without any arguments, which shows that the required argument validation is working. Next, the script is called with the help option, which shows the automatically generated help and usage messages. The robot is then requested to move forward with default and custom duration and speed options.

Going deeper: Functions as first-class objects in Python

In this section, we were able to use `getattr` to look up a function and save it in a variable. This is not possible in all programming languages but is fully supported in Python because functions are treated as first-class objects. This means they can be assigned to variables, placed in lists, or passed as arguments to other functions like any other value. The post "First-class Everything" (http://mng.bz/g7Bl) by the creator of the Python language, Guido van Rossum, is a great read to learn more about this feature in Python. In fact, this feature applies to all objects in Python, not just functions. It makes the language very versatile in terms of how we interact with functions.

5.4.2 Running the script remotely

Now that we have our script in place, we will start calling it from our SSH client. The SSH client can be any computer that is on the same network as the robot. On the client machine, run the following commands to generate SSH keys and transfer them to the robot:

```
ssh-keygen -t rsa
ssh-copy-id robo@robopi
```

The hostname of the Pi was set as `robopi` as part of the installation process documented in appendix B. You can now add a line to your client machine's hosts file with the `robopi` hostname and its associated IP address. In this way, you can follow the examples and connect to the robot using the name `robopi` instead of the robot's IP address. The How-To Geek website provides an excellent guide on how to edit the hosts file on Windows, Mac, and Linux (http://mng.bz/5owz).

At this point, you will be able to execute commands on the robot from the client machine without having to enter a password and with a non-interactive session. What follows is a terminal session run from the client machine:

```
$ ssh robo@robopi whoami
robo
$ ssh robo@robopi '~/pyenv/bin/python --version'
Python 3.9.2
```

In this session, different remote commands are executed to get the name of the current user and the version of the Python interpreter. The Python virtual environment used here is the same as the one created during the installation process covered in appendix B. Next, we can try to execute the robot script. We will be running the script from the standard script location `~/bin` as described in chapter 2:

```
$ ssh robo@robopi '~/pyenv/bin/python ~/bin/cli.py --help'
usage: cli.py [-h] [--duration DURATION] [--speed SPEED] name

robot cli

positional arguments:
  name                  name of movement

optional arguments:
  -h, --help            show this help message and exit
  --duration DURATION   movement duration
  --speed SPEED         movement speed
$ ssh robo@robopi '~/pyenv/bin/python ~/bin/cli.py forward'
calling forward with kwargs {}
```

We have generated the script's help message and requested the robot to move forward remotely over an SSH connection. We can now use the `time` command to measure the execution time when commands are run locally and when they are run remotely over SSH. The `time` command is available on both Mac and Linux. If you are using Windows, you can employ the PowerShell `Measure-Command` instead to measure the execution time. The output first shows the time needed to run the command locally on the robot and then the time needed to establish an SSH connection and execute the command. The timing that we care about is listed under the label `real`. In this session, local execution took 10 ms, while the same command over SSH took 314 ms:

```
$ time whoami
robo
real    0m0.010s
user    0m0.001s
sys     0m0.010s

$ time ssh robo@robopi whoami
robo
real    0m0.314s
user    0m0.084s
sys     0m0.004s
```

The reason for this additional time is that in this approach, a new SSH connection has to be established each time a new command is to be executed. It is useful to make these measurements to know what the overhead of this approach is. The `motor` module has a `noop` function that performs no operation and is perfect for measuring the execution times of purely calling functions in the module:

```
$ time cli.py noop
calling noop with kwargs {}
real    0m0.713s
user    0m0.128s
sys     0m0.066s

$ time ssh robo@robopi '~/pyenv/bin/python ~/bin/cli.py noop'
calling noop with kwargs {}
real    0m1.036s
user    0m0.083s
sys     0m0.005s
```

From the output, we can see that a local call takes 713 ms and a remote one takes 1,036 ms. The difference is 323 ms, which is in line with our previous sampling of SSH overheads. The `time` command is a great way to take quick performance measurements. Later in the chapter, as we improve on these performance numbers, we will explore more accurate ways to measure performance within Python itself.

5.4.3 Creating the client-side script

The next step will be implementing the Python function that runs on the client machine, connects to the robot SSH server, and executes the robot commands. The `check_output` function is imported from the `subprocess` module, which is part of the Python standard library. We can use `check_output` to execute the required SSH client commands:

```
from subprocess import check_output
```

Three constants are then defined. The `SSH_USER` and `SSH_HOST` specify the user and host, respectively, to be used for the SSH connections. The `SSH_CLI_CMD` has the path to the Python interpreter and robot script to be remotely executed on the robot:

```
SSH_USER = 'robo'
SSH_HOST = 'robopi'
SSH_CLI_CMD = '~/pyenv/bin/python ~/bin/cli.py'
```

Next, we define the `call_ssh` that will SSH as the user `user` to the host `host` and execute the provided `remote_cmd` on that remote server:

```
def call_ssh(user, host, remote_cmd):
    cmd = ['ssh', f'{user}@{host}', remote_cmd]
    check_output(cmd)
```

The `remote_robot` receives the name of the robot movement command to perform and executes that movement on the remote robot:

```
def remote_robot(robot_cmd):
    call_ssh(SSH_USER, SSH_HOST, SSH_CLI_CMD + ' ' + robot_cmd)
```

Finally, the `main` function loops through a list of movements and calls `remote_robot` for each movement to perform a demonstration:

```
def main():
    commands = ['forward', 'backward', 'spin_right', 'spin_left']
    for command in commands:
        print('remote robot command:', command)
        remote_robot(command)
```

The full script can be saved as `ssh_client.py` on the client machine and then executed.

Listing 5.2 `ssh_client.py`: Executing remote script over an SSH connection

```
#!/usr/bin/env python3
from subprocess import check_output

SSH_USER = 'robo'
SSH_HOST = 'robopi'
SSH_CLI_CMD = '~/pyenv/bin/python ~/bin/cli.py'

def call_ssh(user, host, remote_cmd):
    cmd = ['ssh', f'{user}@{host}', remote_cmd]
    check_output(cmd)

def remote_robot(robot_cmd):
    call_ssh(SSH_USER, SSH_HOST, SSH_CLI_CMD + ' ' + robot_cmd)
```

```
def main():
    commands = ['forward', 'backward', 'spin_right', 'spin_left']
    for command in commands:
        print('remote robot command:', command)
        remote_robot(command)

main()
```

This script issues remote commands to the robot, making it move forward, backward, and then spin right and left.

5.5 Creating web services for robots

We will now use the HTTP protocol to create web services to control the robot remotely. A web service allows machines to make calls to each other over a network. We need to implement a set of web services in Python that meet the following requirements:

- A set of web services should be created in Python that can be called with a movement action and be given optional speed and duration parameters.
- The web services should use the HTTP methods correctly. Specifically, calls using the GET method should not change the state of the robot, and all movement requests should be processed using the POST method.
- All web service calls should return their data in the JSON format, and any expected input data to the web services should also be encoded in the JSON format.

5.5.1 Creating our first web service

When creating a web application in Python, it is often a good idea to use a web framework. There are many popular options to choose from. We will use the Tornado web framework, as it has many features and can safely interact with the robot hardware. Run the following line on the Pi to install the Tornado Python package:

```
$ ~/pyenv/bin/pip install tornado
```

We can start creating our first web application. This web application will expose a single web service that returns the current time on our robot server. First, we import the required modules. The `datetime` module will let us get the current time on the server. The Tornado `IOLoop` is needed to run the web server. The `RequestHandler` and `Application` objects will help us define the behavior of our web application:

```
from datetime import datetime
from tornado.ioloop import IOLoop
from tornado.web import RequestHandler, Application
```

The next step is to define the `MainHandler` object that will handle incoming requests. We define one method called `get` to handle incoming HTTP GET requests. Each time it is called, it will save the current time as a string and then call the `write` method with the time stamp in a dictionary. In the Tornado framework, whenever you provide the `write` method with a dictionary object, it will automatically convert the output to the

JSON format and set the appropriate HTTP response headers to indicate the content type is JSON:

```
class MainHandler(RequestHandler):
    def get(self):
        stamp = datetime.now().isoformat()
        self.write(dict(stamp=stamp))
```

We then create a Tornado application which will route incoming requests for the root path to `MainHandler`. After that, we set the web server to listen on port 8888 and start the main event loop, which will kick off the web server and handle incoming web requests:

```
app = Application([('/', MainHandler)])
app.listen(8888)
IOLoop.current().start()
```

The full script can be saved as `datews.py` on the Pi and then executed.

Listing 5.3 `datews.py`: Creating web service to report time on the robot server

```
#!/usr/bin/env python3
from datetime import datetime
from tornado.ioloop import IOLoop
from tornado.web import RequestHandler, Application

class MainHandler(RequestHandler):
    def get(self):
        stamp = datetime.now().isoformat()
        self.write(dict(stamp=stamp))

app = Application([('/', MainHandler)])
app.listen(8888)
IOLoop.current().start()
```

Leave the script running in one terminal, and from another terminal connected to the Pi, run the following command to test the new web service:

```
$ curl http://localhost:8888/
{"stamp": "2022-11-27T16:52:36.248068"}
```

The terminal session is using the `curl` command, which is an excellent tool for making HTTP requests in the terminal and viewing their responses. The first call shows the JSON data returned with a time stamp showing the current time on the robot server. We can now run the following command to get more details on the response headers:

```
$ curl -i http://localhost:8888/
HTTP/1.1 200 OK
Server: TornadoServer/6.2
Content-Type: application/json; charset=UTF-8
Date: Sun, 27 Nov 2022 16:52:49 GMT
Etag: "d00b59ccd574e3dc8f86dcadb1d349f53e7711ec"
```

```
Content-Length: 39

{"stamp": "2022-11-27T16:52:49.683872"}
```

This call displays the response headers where we can see that the response content type has been correctly set for the JSON output. You can make these web requests from any web browser on your network by replacing `localhost` with the IP address of the robot on your network.

5.5.2 Creating web services to perform robot movements

We have implemented a simple web service. Now we can upgrade the code to add web services to make the robot move around. We must import two more modules. We will use the `json` module to parse JSON request data. The `motor` module will be used to control the robot motors as in previous chapters:

```
import json
import motor
```

We change the URL pattern to accept any string and then pass that as an argument when calling the method to handle the request. We do this so that the name of the movement to perform can be provided as the URL:

```
app = Application([('/(.*)', MainHandler)])
```

This means we also need to update our previous request method to accept a `name` argument:

```
    def get(self, name):
```

The movement web services will change the state of the robot, so we will perform them when we get POST requests. The `post` method will handle these requests by first reading the request data and parsing it as JSON data. If the web service request has no input data, it will default the value to an empty dictionary. The next step is to take the name of the movement function and retrieve the function from the `motor` module using `getattr`. We can now call the function using the arguments provided in the web service request. The final line of code returns a success status message:

```
    def post(self, name):
        args = json.loads(self.request.body or '{}')
        func = getattr(motor, name)
        func(**args)
        self.write(dict(status='success'))
```

The full script can be saved as `robows.py` on the Pi and then executed.

Listing 5.4 `robows.py`: Creating web service to perform robot movement commands

```
#!/usr/bin/env python3
from datetime import datetime
from tornado.ioloop import IOLoop
```

```
from tornado.web import RequestHandler, Application
import json
import motor

class MainHandler(RequestHandler):
    def get(self, name):
        stamp = datetime.now().isoformat()
        self.write(dict(stamp=stamp))

    def post(self, name):
        args = json.loads(self.request.body or '{}')
        func = getattr(motor, name)
        func(**args)
        self.write(dict(status='success'))

app = Application([('/(.*)', MainHandler)])
app.listen(8888)
IOLoop.current().start()
```

We can again leave the script running in one terminal and run the following commands from another terminal to test the web services:

```
$ curl http://localhost:8888/
{"stamp": "2022-11-27T17:54:30.658154"}

$ curl localhost:8888/
{"stamp": "2022-11-27T17:54:30.658154"}

$ curl -X POST localhost:8888/forward
{"status": "success"}

$ curl -X POST localhost:8888/backward
{"status": "success"}

$ curl -X POST localhost:8888/forward -d '{"speed": 1}'
{"status": "success"}

$ curl -X POST localhost:8888/forward -d '{"duration": 0.5, "speed": 1}'
{"status": "success"}
```

In the terminal session, we first check whether our time web service is still working. The second call demonstrates a shorter way to refer to a URL by not specifying the protocol. We then make a web service call to move the robot forward and then backward. The last two calls show how we can provide custom speed and duration settings for our movements.

5.6 Calling web services from Python

Now that we have these powerful web services in place, we can move on to create code that calls them from anywhere in the network to make the robot move around. We need to implement a web service client in Python that meets the following requirements:

- We should implement a function in Python that will receive the movement name and a set of optional movement arguments and then place the needed HTTP call to the robot web service to execute this movement.

- The implementation should use HTTP persistent connections to have a better network performance by having a lower latency when issuing multiple movement calls.

5.6.1 Using the read–evaluate–print loop to call web services

As a starting step, it will be helpful to start making calls to the web server using the Python REPL (read–evaluate–print loop). In this way, we can explore the different ways we can call web services and the results and data structures they will return. Open a REPL session on the client machine. The first part of our REPL adventure will be importing the modules we need. The command `urlopen` will be used to make calls to the web server, and `json` will be used to parse the JSON responses:

```
>>> from urllib.request import urlopen
>>> import json
```

The next line we execute will connect to the web server and consume the web service that returns the current time on the robot server. The raw JSON response is returned as bytes:

```
>>> urlopen('http://robopi:8888/').read()
b'{"stamp": "2022-11-28T14:30:41.314300"}'
```

We can save this response into a variable and then parse it so that we can access the time stamp value itself:

```
>>> response = urlopen('http://robopi:8888/').read()
>>> response
b'{"stamp": "2022-11-28T14:31:08.859478"}'
>>> json.loads(response)
{'stamp': '2022-11-28T14:31:08.859478'}
>>> result = json.loads(response)
>>> result
{'stamp': '2022-11-28T14:31:08.859478'}
>>> result['stamp']
'2022-11-28T14:31:08.859478'
```

Now, let's move on to calling some web services to make the robot move around. When we provide a value for the `data` argument, `urlopen` will automatically set the HTTP method to be a POST method. The following call will make the robot move forward:

```
>>> urlopen('http://robopi:8888/forward', data=b'').read()
b'{"status": "success"}'
```

We can set custom movement options such as the speed with

```
>>> urlopen('http://robopi:8888/forward', data=b'{"speed": 1}').read()
b'{"status": "success"}'
```

We have now done enough exploring to slap together the first implementation of our web service client.

5.6.2 Creating a basic web service client

The initial version of the client will have everything we need, except persistent connections. We import the same modules as in the REPL session to make requests to the web server and deal with JSON data:

```
from urllib.request import urlopen
import json
```

Next, we define the `ROBO_URL` constant providing the base part of the URLs that we will use to make our calls:

```
ROBO_URL = 'http://robopi:8888/'
```

The `call_api` will place the actual calls to the web service API. It receives the full URL and the requested data as a dictionary. It converts the received data to a JSON format and then calls the `encode` method to convert it to a bytes data type as expected by `urlopen`. Then, `urlopen` is called with the associated URL and request data:

```
def call_api(url, data):
    data = json.dumps(data).encode()
    urlopen(url, data).read()
```

The `call_robot` function receives the movement name and any optional movement arguments. The URL for the related movement is generated, and then `call_api` is called:

```
def call_robot(func, **args):
    call_api(ROBO_URL + func, args)
```

The remaining parts of the script demonstrate the usage of the client by making different calls to `call_robot`:

```
call_robot('forward')
call_robot('backward')
call_robot('forward', duration=0.5, speed=1)
call_robot('backward', duration=0.5, speed=1)
call_robot('spin_right')
call_robot('spin_left')
```

The full script can be saved as `client_basic.py` on the Pi and then executed.

Listing 5.5 `client_basic.py`: Calling a remote web service on the robot from a client

```
#!/usr/bin/env python3
from urllib.request import urlopen
import json

ROBO_URL = 'http://robopi:8888/'

def call_api(url, data):
    data = json.dumps(data).encode()
    urlopen(url, data).read()
```

```
def call_robot(func, **args):
    call_api(ROBO_URL + func, args)

call_robot('forward')
call_robot('backward')
call_robot('forward', duration=0.5, speed=1)
call_robot('backward', duration=0.5, speed=1)
call_robot('spin_right')
call_robot('spin_left')
```

When the script is run, it will make the robot move forward and backward using the default duration and speed settings. Next, forward and backward will be called again with custom settings. Finally, the robot will be made to spin right and left.

Robots in the real world: Robot swarms

Having the ability to communicate with our robots in our software is a powerful and essential feature for a number of robotic applications. Swarm robotics is one of the fields that become possible once you have a mechanism to have robots in your swarm communicate with each other. By using the swarm intelligence or collective behavior of a swarm of robots, we start getting the emergence of intelligent global behavior. This swarm intelligence is found in nature with the sophistication in the design of ant colonies and beehives.

The real-world applications of these robot swarms vary from search-and-rescue missions to different medical applications. The Big Think (http://mng.bz/6nDy) article on this subject shows a nice example of a robot swarm and has a good discussion of the different practical applications of the technology.

5.6.3 Creating a web service client with persistent connections

Now that we have a basic client working, we can upgrade it to have persistent connections to improve the performance of our requests. The approach for this client will be very similar to the previous one but will use a different set of libraries. The first step will be to import the `HTTPConnection` object that offers persistent connection capabilities:

```
from http.client import HTTPConnection
import json
```

The `call_api` function will need to be changed to accept a connection object. After encoding the request body as JSON, we then send the request to the web server using the provided connection object. The request will use a POST method and will make a call to the provided URL with the generated request body. Then, we can use the `getresponse` method to read the response:

```
def call_api(conn, url, data):
    body = json.dumps(data).encode()
    conn. request('POST', url, body)
```

```
    with conn.getresponse() as resp:
        resp.read()
```

The `call_robot` function receives the connection object as an argument and passes the movement name as the requested URL and the movement arguments as the request body:

```
def call_robot(conn, func, **args):
    return call_api(conn, '/' + func, args)
```

We then create an `HTTPConnection` object with the robot hostname and web server port number. A number of calls are then made to `call_robot` to demonstrate its functionality:

```
conn = HTTPConnection('robopi:8888')
for speed in [1, 2, 3]:
    call_robot(conn, 'spin_right', speed=speed)
    call_robot(conn, 'spin_left', speed=speed)
```

The full script can be saved as `client_persist.py` on the Pi and then executed.

Listing 5.6 `client_persist.py`: Using persistent connections to call web services

```
#!/usr/bin/env python3
from http.client import HTTPConnection
import json

def call_api(conn, url, data):
    body = json.dumps(data).encode()
    conn.request('POST', url, body)
    with conn.getresponse() as resp:
        resp.read()

def call_robot(conn, func, **args):
    return call_api(conn, '/' + func, args)

conn = HTTPConnection('robopi:8888')
for speed in [1, 2, 3]:
    call_robot(conn, 'spin_right', speed=speed)
    call_robot(conn, 'spin_left', speed=speed)
```

When the script is run, it will go through three different speed settings and make the robot spin right and left at each setting.

Going deeper: Persistent connections

Under the hood, HTTP requests are transmitted over a TCP connection. Back in the day, each HTTP request would need to go through a new TCP connection. The HTTP protocol was then enhanced to allow for persistent connections or the ability to make multiple requests over a single TCP connection. This improved the network performance of web clients such as web browsers, as it cuts out the overhead of opening

(continued)

a new TCP connection for additional HTTP requests. The Mozilla Foundation documentation on the HTTP protocol (https://developer.mozilla.org/Web/HTTP) covers the topic very well and is an excellent reference for getting more low-level details on the subject.

The performance benefits of using persistent connections make it well worth the effort. It is a standard feature in all modern web browsers and will help us in building time-sensitive real-time robotic applications later in the book.

5.6.4 Measuring client performance

We have gone through all this trouble to add persistent connections. It's worthwhile to create a script to measure the performance of this client. We can use the script to compare the timing of a fresh connection compared to reusing a persistent connection. These timings can also be compared to the results we obtained from the SSH client earlier in the chapter. Finally, we can make a comparison of local web service calls and remote calls over Wi-Fi and wired Ethernet connections.

We will import `mean` to calculate the mean or average of our performance timings and `stdev` to calculate their sample standard deviation. The `perf_counter` function in the `time` module is to record the start and end times of function calls to measure performance. The documentation (https://docs.python.org/3/library/time.html) on `perf_counter` provides guidance on using it when doing performance measurements:

```
from statistics import mean, stdev
import time
```

The `get_noop_timing` function starts by saving the current time using the `perf_counter` function. Then, a call will be made to the `noop` movement function on the robot server. This is a no-operation call that we can use to measure performance between our client and server. Then, we calculate the time elapsed and multiply it by a thousand so that the return value is expressed in milliseconds:

```
def get_noop_timing(conn):
    start = time.perf_counter()
    call_robot(conn, 'noop')
    return (time.perf_counter() - start) * 1000
```

We create a `HTTPConnection` object and make a call to the web server. We do this so that the next calls give more consistent results. Next, we create the connection object that we will use for all our measurements. The measurement for the first web service call is saved in the variable `init` so that we can keep track of how long the initial connection establishment and first call took. Then, we take one hundred timing samples and save them in `stats`. Now we can output the initial, maximum, average, minimum, and standard deviation of samples:

```
conn_initial = HTTPConnection('robopi:8888')
get_noop_timing(conn_initial)
conn = HTTPConnection('robopi:8888')
init = get_noop_timing(conn)
stats = [get_noop_timing(conn) for i in range(100)]
print(' init:', init)
print('  max:', max(stats))
print('  avg:', mean(stats))
print('  min:', min(stats))
print('stdev:', stdev(stats))
```

The full script can be saved as `client_measure.py` on the Pi and then executed.

Listing 5.7 `client_measure.py`: Measuring performance when calling web services

```
#!/usr/bin/env python3
from http.client import HTTPConnection
from statistics import mean, stdev
import time
import json

def call_api(conn, url, data):
    body = json.dumps(data).encode()
    conn.request('POST', url, body)
    with conn.getresponse() as resp:
        resp.read()

def call_robot(conn, func, **args):
    return call_api(conn, '/' + func, args)

def get_noop_timing(conn):
    start = time.perf_counter()
    call_robot(conn, 'noop')
    return (time.perf_counter() - start) * 1000

conn_initial = HTTPConnection('robopi:8888')
get_noop_timing(conn_initial)
conn = HTTPConnection('robopi:8888')
init = get_noop_timing(conn)
stats = [get_noop_timing(conn) for i in range(100)]
print(' init:', init)
print('  max:', max(stats))
print('  avg:', mean(stats))
print('  min:', min(stats))
print('stdev:', stdev(stats))
```

When we run the script, it will collect all the performance measurement timings and output them to the terminal. The following script was run locally on the robot server itself:

```
$ client_measure.py
 init: 2.5157280000485116
  max: 1.9314019999683296
  avg: 1.8538593599976139
  min: 1.812051000001702
stdev: 0.028557077821141714
```

These numbers give us a sense of the overhead in making the web requests end to end, even before having any network packets exit the robot onto the network. Let's look at the numbers we get when we connect to the robot server from a wired Ethernet connection on the network:

```
$ client_measure.py
 init: 4.3936739675700665
  max: 3.5557260271161795
  avg: 2.244193991064094
  min: 1.503808016423136
stdev: 0.5216725173049904
```

Compared to the SSH timings of 1,036 ms, these numbers show a huge difference in performance and overhead between the two approaches. We can also see that the standard deviation has increased, which is expected when moving to a physical network. Next, we measure the timings across a wireless Wi-Fi network:

```
$ client_measure.py
 init: 8.047391020227224
  max: 8.70389404008165
  avg: 4.211111041367985
  min: 3.290054970420897
stdev: 0.8859955886558311
```

These numbers demonstrate that wired network connections can offer better performance then wireless ones. Namely, the initial connection time and average and standard deviation are all better with a wired connection. Standard deviation measures how much variation we have in our measurements. We can see from the standard deviation numbers that performance varies more on a wireless network compared to a wired network. By comparing the timing of initially establishing a connection (8.05 ms) to the average timing on a persistent connection (4.21 ms), we can see we get an almost double performance gain when using persistent connections.

Robots in the real world: Real-time computing

The ability to have low latency communication with our robots makes time-sensitive applications such as real-time computing possible. An example of one of these types of applications is using analog joysticks to control robot movements, which we will do later in the book. This is a very time-sensitive application, and if there is a significant lag between joystick interactions and robot movements, the whole application will fail to function correctly.

Another example is automotive manufacturing where multiple robots are working together on a production line to assemble a car. Different robots will weld, drill, and pass parts to each other. It is critical that these different tasks are performed within set time frames, or the process along the assembly line will be disrupted. This article on real-time systems (http://mng.bz/or1M) covers the topic well in the context of robotics and computer vision.

Summary

- Wi-Fi connectivity gives the robot the greatest freedom of movement.
- Tornado is a feature-rich web framework that can safely interact with the hardware on the robot.
- The `argparse` module is part of the Python standard library and can be used to parse command line arguments.
- The `time` command can be used to measure the execution time when commands are run locally and when they are run remotely over SSH.
- The `json` module is used to parse JSON request data.
- The `urlopen` module can be used to make calls to the web server.
- The use of persistent connections provides significant performance gains.

6 Creating robot web apps

This chapter covers

- Creating a desktop- and mobile-friendly web application to control robots
- Measuring web application performance using web browser tools
- Creating dynamic pages using Tornado templates
- Enabling enhanced web logging to detect web request failures

This chapter will teach you how to build a web application to control your robot. The application will work equally well on desktop computers and mobile phones. The full range of robot movements will be available through the application, along with commands that can be used to measure the network performance of the application, end to end. As we build the application, you will learn useful techniques for measuring the application performance, as well as detecting and fixing certain types of web request failures.

Web applications provide a powerful platform to build a mechanism for controlling the robot by a human operator. Web apps are accessible both from the desktop application and mobile devices. They also work consistently across the main desktop operating systems (i.e., Windows, Mac, and Linux).

6.1 Hardware stack

Figure 6.1 shows the hardware stack, with the specific components used in this chapter highlighted. In this chapter, we can use the mouse as a human-interactive device to interact with the robot through our web app.

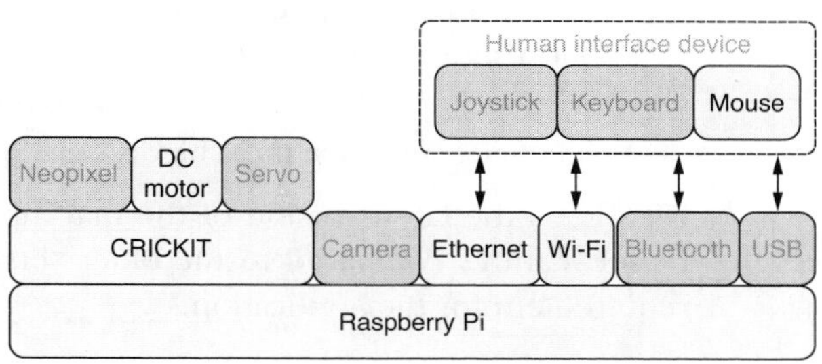

Figure 6.1 Hardware stack: the DC motors will be controlled over the network using a web interface.

The web application can be accessed either through wired networks using the Ethernet port or wireless networks using the Wi-Fi hardware. The best user experience for the mobile access is over the Wi-Fi connection, as it offers full portability. When the web interface is accessed from a desktop, a mouse can be used as the human interface device to control the robot by clicking on the desired robot movement buttons. In later chapters, we will control the robot using a keyboard and a joystick.

6.2 Software stack

Details of the specific software used in this chapter are described in figure 6.2. There are three main applications that will be created here. The first is a basic web application that displays the current time on the robot server (`basic_web`). Then, we will create an application to move the robot forward and backward (`forward_web`). Finally, a mobile-friendly application will be created with a full range of robot movement commands (`full_web`). The Tornado web framework will be used to create these web applications. The built-in template feature of the framework will be used to create dynamic content. The `datetime` and `os` Python modules will be used to calculate the time on the server and read values from environment variables.

Application	`basic_web, forward_web, full_web`
Libraries	Tornado web framework
Python	`datetime, os.path, os`
Linux	TCP/IP network stack
Hardware	Network hardware (Ethernet/Wi-Fi)

Figure 6.2 Software stack: web browsers on the network will connect through a Tornado web application.

6.3 Moving robots forward and backward over the web

The first web application we create will perform the basic forward and backward robot movements. We need to create a web application to meet the following requirements:

- A Python web application that allows users to move the robot forward and backward should be created.
- The web app should use the HTML5 standard.
- The user interface must be desktop and mobile friendly.

HTML5 is the latest version of the markup language used on the web, and it offers richer features compared to the older versions. For this reason, we have put it as a requirement for the application.

6.3.1 Creating a basic web application

Let us take some simple first steps and create a web application that displays the time on the robot web server. The first step is to import all the required modules. From Tornado, we import `IOLoop`, `RequestHandler`, and `Application`, as we have done in previous chapters to set up and run our web application. We then import `enable_pretty_logging` to enable logging output. The `datetime` object will be used to get the current time. The `dirname` function will get the directory name of a path. The `os` module will be used to access environment variables:

```
from tornado.ioloop import IOLoop
from tornado.web import RequestHandler, Application
from tornado.log import enable_pretty_logging
from datetime import datetime
from os.path import dirname
import os
```

The Tornado web framework has a powerful debug mode offering features such as automatic reloading and generated error pages that help both when developing and debugging a web application. The next line of code sets the global variable `DEBUG` to be true or false, depending on whether the environment variable `ROBO_DEBUG` has been defined. In this way, the same code can be used for development or production use, and its debugging behavior can be defined outside the code base through environment variables:

```
DEBUG = bool(os.environ.get('ROBO_DEBUG'))
```

The next step is to set `TEMPLATE_PATH` to the path of the templates directory. This directory will contain the Tornado template files that will be used to generate HTML content. This path is automatically calculated as a subdirectory called `templates` in the same directory as the Python code. Place all HTML template files in this directory:

```
TEMPLATE_PATH = (dirname(__file__) + '/templates')
```

We can now define the `MainHandler` object that will handle incoming requests. It will calculate the current time and save the value as a string in a variable called `stamp`.

Then, the `basic.html` template is rendered with the `stamp` variable and sent to the web browser:

```
class MainHandler(RequestHandler):
    def get(self):
        stamp = datetime.now().isoformat()
        self.render('basic.html', stamp=stamp)
```

The last block of code calls `enable_pretty_logging` to enable logging output and defines `settings` with the application settings. These settings are then provided to `Application`, and the application server is started:

```
enable_pretty_logging()
settings = dict(debug=DEBUG, template_path=TEMPLATE_PATH)
app = Application([('/', MainHandler)], **settings)
app.listen(8888)
IOLoop.current().start()
```

The full script can be saved as `basic_web.py` on the Pi.

Listing 6.1 `basic_web.py`: Web application that displays the time on the robot

```
#!/usr/bin/env python3
from tornado.ioloop import IOLoop
from tornado.web import RequestHandler, Application
from tornado.log import enable_pretty_logging
from datetime import datetime
from os.path import dirname
import os

DEBUG = bool(os.environ.get('ROBO_DEBUG'))
TEMPLATE_PATH = (dirname(__file__) + '/templates')

class MainHandler(RequestHandler):
    def get(self):
        stamp = datetime.now().isoformat()
        self.render('basic.html', stamp=stamp)

enable_pretty_logging()
settings = dict(debug=DEBUG, template_path=TEMPLATE_PATH)
app = Application([('/', MainHandler)], **settings)
app.listen(8888)
IOLoop.current().start()
```

Before we can execute the script, we should create the `basic.html` template. We will run through each part of this template file. The first line of the file is required to declare to the web browser that this file is using HTML5. Then, we have our opening `html` tag, which defines the document language as English:

```
<!DOCTYPE HTML>
<html lang="en">
```

The next part is the `head` section of the HTML document. The website title is provided, and the `meta` tag is then used to set the `viewport` metadata so that the web application will be displayed correctly on both desktop and mobile browsers. Next, the font for the page is set as `Verdana` using the `style` tag:

```
<head>
  <title>Robot Web</title>
  <meta name="viewport" content="width=device-width">
<style>
body {
  font-family: Verdana, sans-serif;
}
</style>
</head>
```

The last part of the template contains the `body` portion of the document. The `h1` tag is used to provide header content, and finally, the `stamp` template variable is then placed under this header to display the current time on the robot:

```
<body>
<h1>Robot Web</h1>
{{ stamp }}
</body>
</html>
```

The template can be saved as `basic.html` on the Pi in the templates directory.

Listing 6.2 `basic.html`: HTML user interface that displays the time on the robot

```
<!DOCTYPE HTML>
<html lang="en">
<head>
  <title>Robot Web</title>
  <meta name="viewport" content="width=device-width">
<style>
body {
  font-family: Verdana, sans-serif;
}
</style>
</head>
<body>
<h1>Robot Web</h1>
{{ stamp }}
</body>
</html>
```

We can now execute `basic_web.py` to run the web server. You can access the web application by visiting the address http://robopi:8888 from a computer on your network. Make sure to update the hosts file on the computer to have an entry for `robopi`, as described in chapter 5. You can also access the web app by replacing `robopi` in the URL with the IP address of your robot. When accessing the web app from a mobile device, using the IP address will be an easier option.

When you access the web application, it will display the current time on the robot server. Refresh the page to see the updated time and confirm that the server is responding to multiple requests. Figure 6.3 shows what the application will look like.

Figure 6.3 `basic_web`**: it displays the current time on the robot server.**

6.3.2 Detecting failed requests

Because we have called `enable_pretty_logging`, we get the benefit of outputting the server logs to the terminal. We can inspect this output to see all the incoming requests to the web server and the given responses. Try to access the web application again from multiple browsers or computers. What follows is the log output after the application has been accessed from multiple browsers and computers:

```
$ basic_web.py
[I 221213 17:20:27 web:2271] 200 GET / (10.0.0.30) 7.99ms
[W 221213 17:20:27 web:2271] 404 GET /favicon.ico (10.0.0.30) 1.27ms
[I 221213 17:20:33 web:2271] 200 GET / (10.0.0.30) 2.21ms
[W 221213 17:20:34 web:2271] 404 GET /favicon.ico (10.0.0.30) 1.84ms
[I 221213 17:20:35 web:2271] 200 GET / (10.0.0.30) 1.98ms
[I 221213 17:20:35 web:2271] 200 GET / (10.0.0.30) 2.23ms
[W 221213 17:23:51 web:2271] 404 GET /favicon.ico (10.0.0.15) 1.82ms
[I 221213 17:23:53 web:2271] 200 GET / (10.0.0.15) 2.36ms
[I 221213 17:23:54 web:2271] 200 GET / (10.0.0.15) 2.32ms
[I 221213 17:23:55 web:2271] 200 GET / (10.0.0.15) 2.23ms
```

We can see that after the page is loaded, the browser tries to fetch a file called `favicon.ico` and fails with a "404 not found" HTTP error. This is because we haven't defined a proper way to handle these requests, so they are failing. In the next upgrade, we can address this problem and check the server logs after the change to confirm that the situation has been resolved. This log output is also a great way to see how long it takes for Tornado to provide a response, as the response times also appear in the log output.

6.3.3 Moving robots forward with web apps

Now we will add the forward and backward movements to our application. We will once again import the `motor` module to control the robot's movements:

```
import motor
```

The `Application` object will be enhanced to handle different request URLs and parse the action from the path being requested. The regular expression `/([a-z_]*)` is used to match a path consisting of lowercase letters and the underscore character. This pattern will match all the available movement commands:

```
app = Application([('/([a-z_]*)', MainHandler)], **settings)
```

We now update the `get` method to receive the `name` argument and to render the `forward.html` template:

```
def get(self, name):
    stamp = datetime.now().isoformat()
    self.render('forward.html', stamp=stamp)
```

As done in previous chapters, we will only process movement commands when they come in as `post` requests. The `post` method will then check the value of the `name` variable and call the related movement function. It will then redirect the browser to the web application home page using the `redirect` method:

```
def post(self, name):
    if name == 'forward':
        motor.forward()
    if name == 'backward':
        motor.backward()
    self.redirect('/')
```

The full script can be saved as `forward_web.py` on the Pi.

Listing 6.3 `forward_web.py`: Web application to move the robot forward and backward

```
#!/usr/bin/env python3
from tornado.ioloop import IOLoop
from tornado.web import RequestHandler, Application
from tornado.log import enable_pretty_logging
from datetime import datetime
from os.path import dirname
import os
import motor

DEBUG = bool(os.environ.get('ROBO_DEBUG'))
TEMPLATE_PATH = (dirname(__file__) + '/templates')

class MainHandler(RequestHandler):
    def get(self, name):
        stamp = datetime.now().isoformat()
        self.render('forward.html', stamp=stamp)

    def post(self, name):
        if name == 'forward':
            motor.forward()
        if name == 'backward':
            motor.backward()
        self.redirect('/')

enable_pretty_logging()
settings = dict(debug=DEBUG, template_path=TEMPLATE_PATH)
app = Application([('/([a-z_]*)', MainHandler)], **settings)
app.listen(8888)
IOLoop.current().start()
```

We can now upgrade the HTML template. To resolve the `favicon` problem, we use the following HTML in the `head` portion of our template. This sets no icon for the page and thus instructs the web browser not to fetch the `favicon` file from the web server:

```
<link rel="icon" href="data:,">
```

In the main body of the document, we add two forms. Each form will submit its data, using the `post` method, to either the `forward` path or the `backward` path. The submit buttons for each form have a label that matches the movement action for that form:

```
<form method="post" action="forward">
  <input type="submit" value="Forward">
</form>
<form method="post" action="backward">
  <input type="submit" value="Backward">
</form>
```

The template can be saved as `forward.html` on the Pi in the templates directory.

Listing 6.4 `forward.html`: HTML to move the robot forward and backward

```
<!DOCTYPE HTML>
<html lang="en">
<head>
  <title>Robot Web</title>
  <meta name="viewport" content="width=device-width">
  <link rel="icon" href="data:,">
<style>
body {
  font-family: Verdana, sans-serif;
}
</style>
</head>
<body>
<h1>Robot Web</h1>
{{ stamp }}
<form method="post" action="forward">
  <input type="submit" value="Forward">
</form>
<form method="post" action="backward">
  <input type="submit" value="Backward">
</form>

</body>
</html>
```

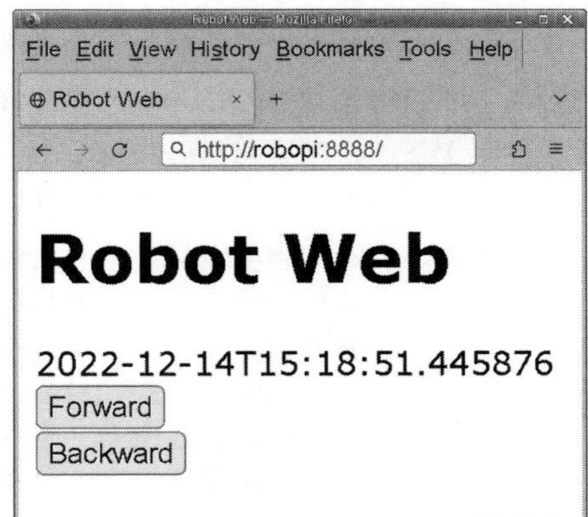

Figure 6.4 `forward_web`: it provides buttons to move the robot forward and backward.

We can now execute `forward_web.py` to run the web server. When you access the web application, press the forward and backward buttons to move the robot backward and forward. Figure 6.4 shows what the application will look like now with these new buttons. We can inspect the log output from a session using the application:

```
$ forward_web.py
[I 221213 17:37:29 web:2271] 200 GET / (10.0.0.30) 7.99ms
[I 221213 17:37:34 web:2271] 302 POST /forward (10.0.0.30) 222.82ms
[I 221213 17:37:34 web:2271] 200 GET / (10.0.0.30) 2.28ms
[I 221213 17:37:35 web:2271] 302 POST /backward (10.0.0.30) 223.56ms
[I 221213 17:37:35 web:2271] 200 GET / (10.0.0.30) 2.25ms
[I 221213 17:37:36 web:2271] 302 POST /backward (10.0.0.30) 224.18ms
[I 221213 17:37:36 web:2271] 200 GET / (10.0.0.30) 2.22ms
```

In the log output, we can see the robot was moved forward once and then backward twice. From the logs, we can see that it usually takes 2 ms to render the main page. It takes around 224 ms to perform a robot movement. The default duration set for a robot movement in the `motor` module is 200 ms. Therefore, these numbers are what we would expect. Finally, we can see the "favicon not found errors" have also been resolved, as they are no longer appearing in the request logs.

Going Deeper: HTML5

HTML5 is the latest version of the markup language used on the web. The standard is maintained by the Web Hypertext Application Technology Working Group (WHATWG), which is a consortium of major browser vendors (Apple, Google, Mozilla, and Microsoft). The HTML Living Standard (https://w3.org/TR/html5) provides full details on the HTML elements and syntax. It is a comprehensive reference on the standard.

HTML forms are used heavily in this chapter to submit the desired movement actions to the robot server. The Mozilla guide on web forms (https://developer.mozilla.org/Learn/Forms) is an excellent resource to explore how forms with different submission options can be created, as well as different input elements contained within the form itself.

6.4 Creating a full-movement web app

We can now move on to create a web application that can call all the robot movement functions. We need to create a web application to meet the following requirements:

- A Python web application that allows users to move the robot forward, backward, right, left, and to spin in both directions should be created.
- A button to call the no-operation `noop` function should be created so that performance measurements can be made within the web browser.
- The buttons in the user interface should use a layout that comfortably supports both mobile touch and desktop mouse interactions.

6.4.1 Creating the full-movement application

On the Python side, we are almost there. We can make some small modifications to our previous application to enable all the movement functions. We will make a small modification to the `get` method so that it uses our new `full.html` template:

```
    def get(self, name):
        stamp = datetime.now().isoformat()
        self.render('full.html', stamp=stamp)
```

The `post` method will now be enhanced to look up the required movement function from the `motor` module and then call the function. After that, we will be redirected to the application home screen:

```
    def post(self, name):
        func = getattr(motor, name)
        func()
        self.redirect('/')
```

The full script can be saved as `full_web.py` on the Pi.

Listing 6.5 `full_web.py`: Web application that supports all robot movement actions

```
#!/usr/bin/env python3
from tornado.ioloop import IOLoop
from tornado.web import RequestHandler, Application
from tornado.log import enable_pretty_logging
from datetime import datetime
from os.path import dirname
import os
import motor

DEBUG = bool(os.environ.get('ROBO_DEBUG'))
TEMPLATE_PATH = (dirname(__file__) + '/templates')

class MainHandler(RequestHandler):
    def get(self, name):
        stamp = datetime.now().isoformat()
        self.render('full.html', stamp=stamp)

    def post(self, name):
        func = getattr(motor, name)
        func()
        self.redirect('/')

enable_pretty_logging()
settings = dict(debug=DEBUG, template_path=TEMPLATE_PATH)
app = Application([('/([a-z_]*)', MainHandler)], **settings)
app.listen(8888)
IOLoop.current().start()
```

The next step will be to upgrade the HTML template. The contents in the `body` will have a number of enhancements made. The text in the title of the page will be given a link so that users can click on the page title to reload the page. A button for each movement function is placed on the screen. The layout of the screen is designed so that similar actions are grouped together in the same row. The first row has the forward and backward buttons. The second row has the buttons to turn left and right. The third row shows buttons to spin left and right. The final row presents the no-operation button.

The buttons use HTML5-named character references so that the buttons have a graphical indication of what they do:

```
<body>
<h1><a href='/'>Robot Web</a></h1>
{{ stamp }}<br><br>
<form method="post">
  <button formaction="forward">&blacktriangle;</button>
  <button formaction="backward">&blacktriangledown;</button>
  <br><br>
  <button formaction="left">&blacktriangleleft;</button>
  <button formaction="right">&blacktriangleright;</button>
  <br><br>
  <button formaction="spin_left">&circlearrowleft;</button>
  <button formaction="spin_right">&circlearrowright;</button>
  <br><br>
  <button formaction="noop">X</button>
</form>
</body>
```

Now we can style the content on the page by updating the style tag. We center the content in the page and remove the default underline text decoration style from links on the page. Next, we move on to style the buttons on the page. Their font size is made three times larger, and a good amount of margin is added to provide a healthy amount of spacing between the buttons. This layout and spacing make it easier to press the buttons with your finger on a touch interface, as the buttons aren't bunched up next to each other. Finally, all the buttons are given the same height and width of 60 px to create a uniform look:

```
<style>
body, a {
  font-family: Verdana, Arial, sans-serif;
  text-align: center;
  text-decoration: none;
}
button {
  font-size: 300%;
  margin: 0px 10px;
  height: 60px;
  width: 60px;
}
</style>
```

The template can be saved as `full.html` on the Pi in the templates directory.

Listing 6.6 `full.html`: HTML user interface that supports all robot movement actions

```
<!DOCTYPE HTML>
<html lang="en">
<head>
  <title>Robot Web</title>
  <meta name="viewport" content="width=device-width">
  <link rel="icon" href="data:,">
```

```
<style>
body, a {
  font-family: Verdana, Arial, sans-serif;
  text-align: center;
  text-decoration: none;
}
button {
  font-size: 300%;
  margin: 0px 10px;
  height: 60px;
  width: 60px;
}
</style>
</head>
<body>
<h1><a href='/'>Robot Web</a></h1>
{{ stamp }}<br><br>
<form method="post">
  <button formaction="forward">&blacktriangle;</button>
  <button formaction="backward">&blacktriangledown;</button>
  <br><br>
  <button formaction="left">&blacktriangleleft;</button>
  <button formaction="right">&blacktriangleright;</button>
  <br><br>
  <button formaction="spin_left">&circlearrowleft;</button>
  <button formaction="spin_right">&circlearrowright;</button>
  <br><br>
  <button formaction="noop">X</button>
</form>
</body>
</html>
```

We can now execute `full_web.py` to run the web server. When you access the web application, press different movement buttons to make the robot perform each available movement. What follows is the log output from a session using the application:

```
$ full_web.py
[I 221214 15:37:32 web:2271] 200 GET / (10.0.0.30) 4.75ms
[I 221214 15:37:34 web:2271] 302 POST /forward (10.0.0.30) 223.77ms
[I 221214 15:37:34 web:2271] 200 GET / (10.0.0.30) 5.26ms
[I 221214 15:37:35 web:2271] 302 POST /backward (10.0.0.30) 223.29ms
[I 221214 15:37:35 web:2271] 200 GET / (10.0.0.30) 4.77ms
[I 221214 15:37:35 web:2271] 302 POST /left (10.0.0.30) 222.85ms
[I 221214 15:37:35 web:2271] 200 GET / (10.0.0.30) 4.78ms
[I 221214 15:37:36 web:2271] 302 POST /right (10.0.0.30) 222.96ms
[I 221214 15:37:36 web:2271] 200 GET / (10.0.0.30) 4.81ms
[I 221214 15:37:40 web:2271] 302 POST /spin_left (10.0.0.30) 223.67ms
[I 221214 15:37:40 web:2271] 200 GET / (10.0.0.30) 4.80ms
[I 221214 15:37:41 web:2271] 302 POST /spin_right (10.0.0.30) 223.42ms
[I 221214 15:37:41 web:2271] 200 GET / (10.0.0.30) 4.84ms
[I 221214 15:37:41 web:2271] 302 POST /noop (10.0.0.30) 1.83ms
[I 221214 15:37:41 web:2271] 200 GET / (10.0.0.30) 4.87ms
```

From the logs, we can see each of the different movement functions being successfully executed and then the main page loading after each function is called. It is also of note that the call to the `noop` page loads in around 2 ms, which indicates a good level of performance.

6.4.2 Web interface design

Figure 6.5 shows how the web application will appear for desktop and laptop users. The icons used for each button reflect the movement that will be performed. Thus, arrows pointing forward and backward indicate the forward and backward robot movements. The right, left, and spin functions have similar graphical representations. The no-operation action is presented with an X to indicate that it will not result in a movement.

Figure 6.6 shows how the application will appear on a mobile device. The buttons use larger fonts and are given a good amount of width and height so that they are large enough to be comfortably pressed on a touch screen. The generous spacing between the buttons also prevents their crowding. The whole application fits on one screen so that all functions can be accessed without the need to scroll.

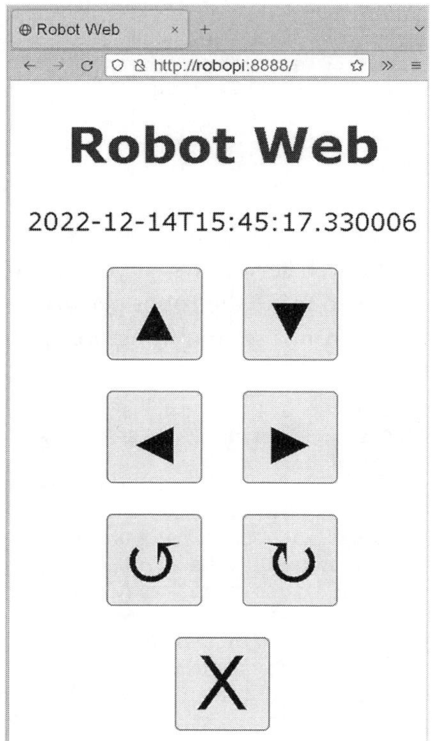

Figure 6.5 Desktop interface: each robot movement can be called by its related button.

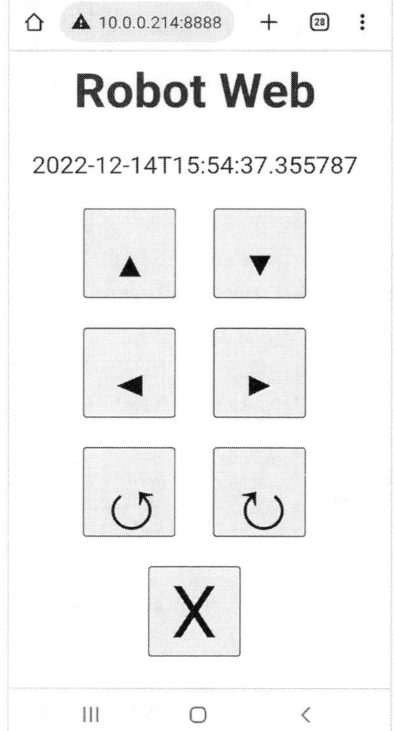

Figure 6.6 Mobile interface: the spacing between the buttons makes touch interaction easier.

Robots in the real world: Controlling robots with web applications

Controlling robots with a web application can provide more convenient access than using a desktop application, as they can be accessed by mobile devices. Operating robots on a factory floor by accessing and controlling the robot from a smartphone gives a greater freedom of movement to the operator than lugging around a laptop.

When needed, they also support the access from devices with larger screens, such as tablet computers or laptops. Different user interfaces can be designed to better support devices with larger screens. In this way, users get the best of both worlds.

Modern web applications are also very extensible. We'll see in the later chapters how to add video streaming from the robot camera to our web applications. This gives the robot operator the ability to see exactly what the robot sees.

6.4.3 Measuring application performance in the browser

Figure 6.7 shows how to measure web application performance in the browser. Both the Firefox and Google Chrome web browsers have a built-in function called Developer Tools, which provides many rich features, such as measuring application performance. Once the tool is accessed in the browser, click on the Network tab. Figure 6.7

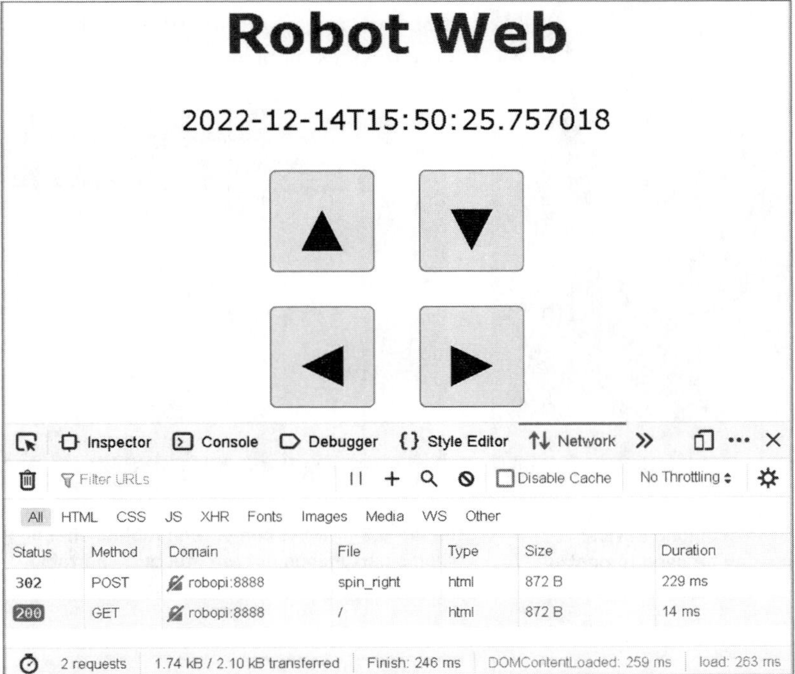

Figure 6.7 Measuring performance: the load times of a page can be measured in the browser itself.

shows a sample measurement taken when the robot was requested to spin right. From the measurements, we can see that it took 229 ms to call the spin right action and then 14 ms to redirect and load the main page. These figures match the ones that we could see on the server from the log output in the terminal. This tool can be very handy when trying to diagnose performance problems with web applications.

6.4.4 Web hardware devices

Figure 6.8 shows how the web application will appear on a smartphone running the iOS operating system. Smartphones offer the greatest level of portability, as they are small enough to fit in your pocket. The downside of using them is their smaller screen, and we are also limited to only using a touch screen interface for interacting with the application.

Figure 6.9 shows the application running on a tablet computer. These types of hardware devices offer much larger screens, which allows us to put many more controls in our user interface.

It is often a good idea to try out your web application on a variety of devices with different screen sizes and web browsers so that you can discover any problems that might occur on specific devices or browsers.

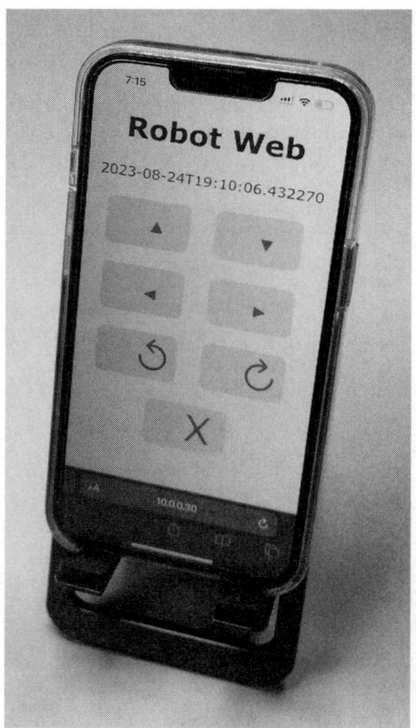

Figure 6.8 Smartphone device: smartphones can be used to control the robot.

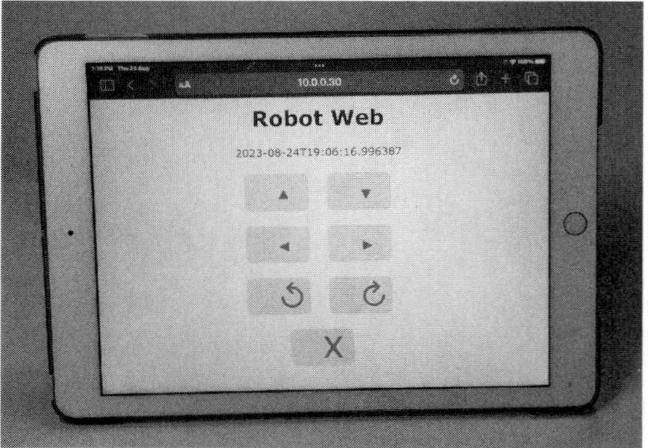

Figure 6.9 Tablet computers: tablets offer bigger screens then smartphones but are still quite portable.

Summary

- Accessing the robot over Wi-Fi gives the best user experience, as both the robot and mobile phone have full portability.
- The built-in template feature of the Tornado web framework is used to create dynamic content in the application.
- One of the benefits of outputting the server logs is that we can see all the incoming requests to the web server and the given responses.
- The buttons use larger fonts and are given a good amount of width and height so that they are large enough to be comfortably pressed on a touch screen.
- Both the Firefox and Google Chrome web browsers have a built-in function called Developer Tools, which provides many rich features, such as measuring application performance.

7 Joystick-controlled robots

This chapter covers

- Reading joystick data using Pygame
- Reading and parsing raw joystick event data
- Measuring the rate of joystick events
- Creating a remote joystick robot controller

Joysticks are one of the most powerful input devices. When it comes to controlling robot motors, they offer much more superior control compared to keyboards and mice. The scenarios covered in this chapter will help you create a fully functional joystick-controlled robot. This chapter will teach you multiple ways of reading events from joystick hardware. We can then create our own event handlers that will perform different robot movements based on specific joystick movements. Along the way, we will also learn how to measure the number of joystick events triggered per second and optimize our code so that it prevents the robot motors from getting flooded with movement requests. Finally, we end the chapter by creating an application that moves the robot using a joystick over the network.

Joystick-controlled robots have a wide array of applications, ranging from remotely operating heavy vehicles on factory floors to performing delicate medical procedures using robotic arms. In the case of robot-assisted surgery, by controlling very small robotic arms, the doctor can perform surgical procedures that would not be possible otherwise.

7.1 Hardware stack

Figure 7.1 shows the hardware stack, with the specific components used in this chapter highlighted. The joystick hardware can be connected to the Raspberry Pi using either a wired USB or a wireless Bluetooth connection. The joystick can also be connected to a remote computer, and the robot movement requests will be transmitted over the network using either Wi-Fi or Ethernet.

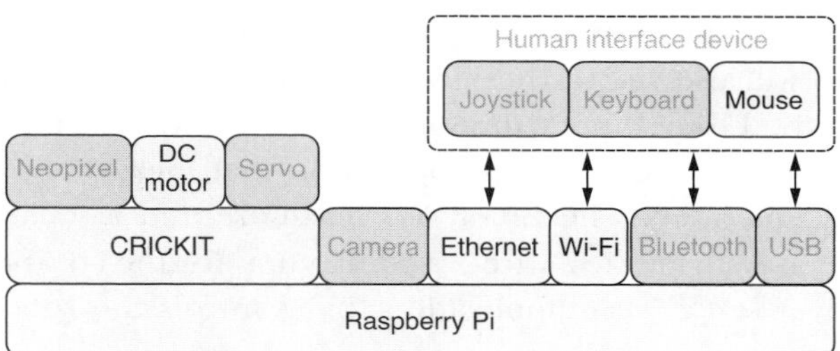

Figure 7.1 Hardware stack: the Joystick will be used to control the robot movements.

The Joystick hardware to be used for this chapter is either the Sony PlayStation 4/5 controller or an Xbox controller. Figure 7.2 shows a photo of a PlayStation 4 controller, and figure 7.3 shows a photo of an Xbox controller. Make sure to check the hardware purchasing guide in appendix A before buying the hardware needed in this chapter.

Figure 7.2 PlayStation 4 controller: this controller is widely available and has good Linux support.

Figure 7.3 Xbox controller: this controller is like the PlayStation controller, but it has two analog sticks.

When connecting the controller to the Raspberry Pi over USB, all you need to do is connect the USB cable between the controller and the Raspberry Pi. No additional software or configuration is required. The Sony PlayStation controller supports a Bluetooth connection. To make use of it, you must first follow the instructions for your controller to put it into a pairing mode. Then you can search and pair the controller like any other Bluetooth device using the graphical Bluetooth application that comes with the Raspberry Pi OS. The final application in the chapter also supports connecting the controller to a remote Linux computer on the same network. On that computer, the same USB and Bluetooth options can be used.

7.2 Software stack

Details of the specific software used in this chapter are provided in figure 7.4. The first few applications will use the Pygame library, as it is a great starting point to work with joystick devices. Then, we will use the `struct` Python module to directly read and parse joystick events from the Linux input subsystem. The `sys` and `time` modules will be used when we create the `joystick_levels` application that measures the rate at which joystick events are generated. The Bluetooth hardware and associated Bluetooth Linux drivers will be used to create a wireless connection for the controller. The chapter ends with the `joystick_remote` application that controls the robot motors using the joystick hardware.

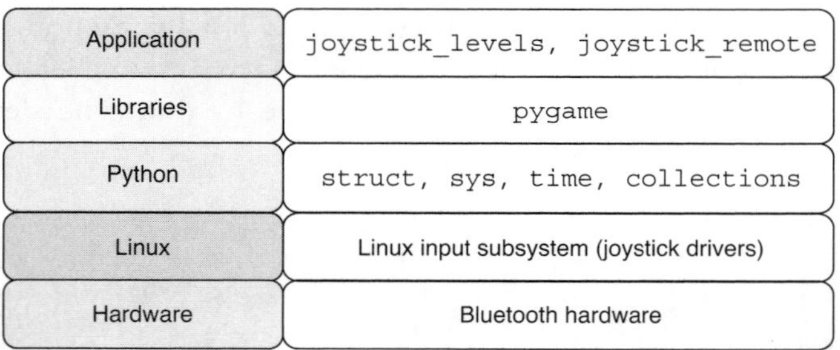

Figure 7.4 Software stack: the Linux input subsystem will be used to read joystick events.

7.3 Joystick events

Figure 7.5 shows the specific joystick events we are most interested in. There are many buttons and sticks on the controller, and each can send events to the connected computer when they are pressed or moved. For the applications in this chapter, we are most interested in the events related to the two analog sticks on the controller. There is one stick for the left and another for the right hand. We will control the robot's movements

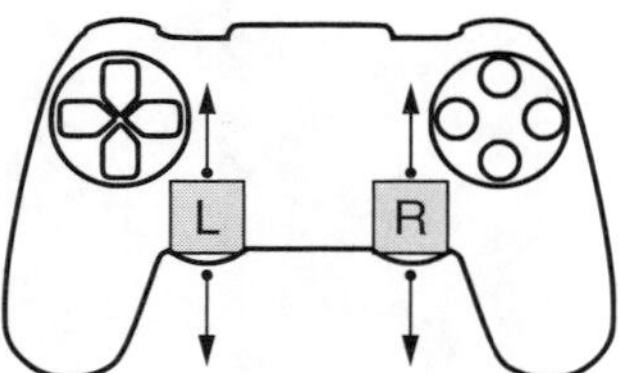

Figure 7.5 Joystick events: the sticks generate *y*-axis events when moved up and down.

by having the throttle on each motor set based on the position of the stick. If the right stick is pushed fully forward, the right motor will be given full throttle power in the forward direction. If the right stick is pulled all the way back, the right motor will be given full throttle power in the backward direction. The same will be done for the left stick and left motor. We will also set the throttle speed or level based on how far each stick is pushed forward or backward. In this way, you can use the joystick to control forward, backward, turn, and spin movements. You can also perform these movements at a slower or faster speed, depending on how far you push the sticks. When the sticks are moved, the stick axis and position are provided as an event. Each stick has the *y*- and *x*-axes. Changes in the vertical position of a stick relate to a *y*-axis event, and changes in the horizontal position of the stick relate to an *x*-axis event.

7.4 Reading joystick events using Pygame

Pygame is a very popular Python module used for writing video games. It has built-in support for reading joystick events and is an excellent starting point for working with joysticks in Python. We need to create an application to meet the following requirements:

- It is necessary to create a Python application that uses the Pygame library to read joystick events.
- We should create an event-handler function in the application that will get called every time there is a stick movement or button-press event.

7.4.1 Detecting events in Pygame

This first program will have an event loop that reads all the events detected and prints them out. Once we have it in place, we can move on to the next section that will focus on joystick events. Run the following line to install the Pygame Python package in our virtual environment:

```
$ ~/pyenv/bin/pip install pygame
```

The first part of our application will import the `pygame` module:

```
import pygame
```

When we run the main event loop, we will need to set how frequently the loop will check for events. This rate is called the frame rate, and we set it in a variable called `FRAME_RATE`. It is set at 60 frames per second, which is a common value for creating a responsive application. If this value is too small, the application will not be very responsive, and if too high, it would put an unnecessary load on the computer without providing an improved user experience. Human beings cannot perceive frame rates beyond 60 frames per second. We save the window height and width in the variable called `WINDOW_SIZE`. The size of the window isn't too important in our application because we won't be drawing anything in the window:

```
FRAME_RATE = 60
WINDOW_SIZE = [100, 100]
```

We now define the function `main` that is at the heart of our program. We call the `pygame.init` function to initialize the Pygame module. Then, we create a window called `screen`. We then create a `Clock` object that will be used in our event loop to process events at the desired frame rate. The next block is the main event loop, which is run constantly until the application is exited. The subsequent available event is fetched by calling `pygame.event.get`. The details of this event are then printed. The type of event is checked to see whether it is a `pygame.QUIT` type of event. If so, the application is exited by calling `pygame.quit`, and then we return from the main function. Finally, the last line of the loop calls `clock.tick` with the configured frame rate:

```
def main():
    pygame.init()
    screen = pygame.display.set_mode(WINDOW_SIZE)
    clock = pygame.time.Clock()
    while True:
        for event in pygame.event.get():
            print('event detected:', event)
            if event.type == pygame.QUIT:
                pygame.quit()
                return
        clock.tick(FRAME_RATE)
```

The last line of the application calls the `main` function:

```
main()
```

The full script can be saved as `pygame_events.py` on the Pi and then executed.

Listing 7.1 `pygame_events.py`: Using the Pygame library to print joystick events

```
#!/usr/bin/env python3
import pygame

FRAME_RATE = 60
WINDOW_SIZE = [100, 100]

def main():
    pygame.init()
    screen = pygame.display.set_mode(WINDOW_SIZE)
    clock = pygame.time.Clock()
    while True:
        for event in pygame.event.get():
            print('event detected:', event)
            if event.type == pygame.QUIT:
                pygame.quit()
                return
        clock.tick(FRAME_RATE)

main()
```

The application requires a graphical environment to run as it creates windows. You can run it on the desktop environment on the Pi directly or remotely over a VNC session. The output is taken from a session of running the command. In the session that follows, the application was started, and the letter A was pressed down and released on the keyboard. The mouse was moved around the application, and then the window was closed. The keyboard, mouse, and window close events can be seen in the output as being detected:

```
$ pygame_events.py
pygame 2.1.2 (SDL 2.0.14, Python 3.9.2)
Hello from the pygame community. https://www.pygame.org/contribute.html
event detected: <Event(32774-WindowShown {'window': None})>
event detected: <Event(32777-WindowMoved {'x': 464, 'y': 364, 'window': ...
event detected: <Event(32770-VideoExpose {})>
event detected: <Event(32776-WindowExposed {'window': None})>
event detected: <Event(32768-ActiveEvent {'gain': 1, 'state': 1})>
event detected: <Event(32785-WindowFocusGained {'window': None})>
event detected: <Event(32788-WindowTakeFocus {'window': None})>
event detected: <Event(768-KeyDown {'unicode': 'a', 'key': 97, 'mod': ...
event detected: <Event(771-TextInput {'text': 'a', 'window': None})>
event detected: <Event(769-KeyUp {'unicode': 'a', 'key': 97, 'mod': 0, ...
event detected: <Event(32768-ActiveEvent {'gain': 1, 'state': 0})>
event detected: <Event(32783-WindowEnter {'window': None})>
event detected: <Event(1024-MouseMotion {'pos': (99, 2), 'rel': (0, 0), ...
event detected: <Event(32768-ActiveEvent {'gain': 0, 'state': 0})>
event detected: <Event(32784-WindowLeave {'window': None})>
event detected: <Event(32788-WindowTakeFocus {'window': None})>
event detected: <Event(32787-WindowClose {'window': None})>
event detected: <Event(256-Quit {})>
```

Going deeper: Frame rate

Frame rate is often measured in frames per second. It is a very important aspect of human interaction with computers. In this chapter, we focus on creating applications that read joystick events and react to them fast enough to create real-time applications. If our frame rate drops to a very low level, it will be visible to our robot operator. There will be a noticeable lag in our application reacting to our actions.

Even though we have set the frame rate in the initial application to 60 frames per second, a lower frame rate of 30 frames per second is still popular and comfortable for many applications. In the coming chapters, the default frame capture rate of the Pi camera is 30 frames per second. Images displayed at this rate will appear as a smooth video feed. As we perform demanding tasks like face detection and frame rates drop to much lower levels, it will become very noticeable and disruptive. Thus, we will solve these problems as we encounter them through software optimization.

Whether we deal with video playback, joystick events, or any other highly interactive user application, it will often boil down to measuring the frame rates and making sure the software design maintains the target frame rate so that the user experience is not affected.

7.4.2 Detecting joystick events

We can now add the capability to detect and handle joystick events to our application. The following two lines are added to the `main` function. The first line calls `Joystick` to set up the controller device object and saves it in the `joystick` variable. We then output the name of the joystick controller device:

```
joystick = pygame.joystick.Joystick(0)
print('joystick name:', joystick.get_name())
```

One line is added to our previous event loop to call the `handle_event` function each time we detect a new event:

```
while True:
    for event in pygame.event.get():
        if event.type == pygame.QUIT:
            pygame.quit()
            return
        handle_event(event)
    clock.tick(FRAME_RATE)
```

The event-handler function can now be defined. It will focus only on joystick events and print a different message when a button is pressed or when one of the sticks on the controller is moved around:

```
def handle_event(event):
    if event.type == pygame.JOYBUTTONDOWN:
        print('button pressed', event.button)
    if event.type == pygame.JOYAXISMOTION:
        print('axis motion', event.axis, event.value)
```

The full script can be saved as `pygame_joystick.py` on the Pi and then executed.

Listing 7.2 `pygame_joystick.py`: Detecting specific joystick events with Pygame

```
#!/usr/bin/env python3
import pygame

FRAME_RATE = 60
WINDOW_SIZE = [100, 100]

def handle_event(event):
    if event.type == pygame.JOYBUTTONDOWN:
        print('button pressed', event.button)
    if event.type == pygame.JOYAXISMOTION:
        print('axis motion', event.axis, event.value)

def main():
    pygame.init()
    screen = pygame.display.set_mode(WINDOW_SIZE)
    clock = pygame.time.Clock()
    joystick = pygame.joystick.Joystick(0)
    print('joystick name:', joystick.get_name())
```

```
    while True:
        for event in pygame.event.get():
            if event.type == pygame.QUIT:
                pygame.quit()
                return
            handle_event(event)
        clock.tick(FRAME_RATE)

main()
```

In the session that follows, the application was started, four different buttons were pressed, and the stick was moved around to different positions. We can see that each button has a unique identifier, and the stick movements have details to identify the axis of movement and the position the stick was moved to:

```
$ pygame_events.py
pygame 2.1.2 (SDL 2.0.14, Python 3.9.2)
pygame 2.1.2 (SDL 2.0.14, Python 3.9.2)
Hello from the pygame community. https://www.pygame.org/contribute.html
joystick name: Sony Interactive Entertainment Wireless Controller
button pressed 0
button pressed 3
button pressed 2
button pressed 1
axis motion 0 -0.003936887722403638
axis motion 0 -0.003936887722403638
axis motion 4 0.003906369212927641
axis motion 4 -0.027466658528397473
```

In the following sections, we will learn how to read the event data in more detail.

7.5 Reading Linux joystick events

Using Pygame gave us a good introduction into interacting with joystick events in Python and creating a graphical application, which we will be doing more in the coming chapters. However, it is a library focused more on creating video games, so it is not perfectly suitable for our needs. Luckily, on Linux, there is very good support for joystick devices, and their event data can be directly read by Python applications. The benefit of doing this is that we can avoid the overhead and complexity of running a whole video game engine and can instead focus purely on the task at hand of reading joystick events. We will parse the joystick event data and focus on handling joystick events on the *y*-axis, which are the stick movements we are most interested in for our robot application. We need to create an application to meet the following requirements:

- The Python application should directly read Linux joystick events.
- It should differentiate whether the event was a button-press or a stick movement event.
- It should filter axis events to handle only *y*-axis events on both sticks.
- It should be able to calculate the direction and percentage of movement on the *y*-axis.

7.5.1 Exploring the Linux input subsystem

To read the joystick event data, we first need to explore the Linux input subsystem. The documentation (https://www.kernel.org/doc/html/latest/input/) is very comprehensive and will be the basis for our exploration and implementation. Everything we are interested in is covered in the "Linux joystick support" chapter. From the documentation, we can see that each connected joystick is exposed as a device file on the filesystem. This is a common approach on Unix systems. The joystick devices are automatically created when joysticks are connected, and a common naming convention is followed. This makes it easy to list them. The following terminal session shows how we can list the joystick devices on a system:

```
$ ls /dev/input/js*
/dev/input/js0
```

From the output, we can see that one joystick is connected. If additional joysticks were connected, they would be called `/dev/input/js1` and `/dev/input/js2`. The documentation also covers the `jstest` command that can be used to connect to the joystick in the terminal and see details of the generated joystick events. Run the following line to install the command:

```
$ sudo apt install joystick
```

We can now run `jstest` to obtain a live view of the joystick events in our terminal:

```
$ jstest /dev/input/js0
Driver version is 2.1.0.
Joystick (Wireless Controller)
  has 8 axes (X, Y, Z, Rx, Ry, Rz, Hat0X, Hat0Y)
and 13 buttons (BtnA, BtnB, BtnX, BtnY, BtnTL, BtnTR, BtnTL2, BtnTR2,
  BtnSelect, BtnStart, BtnMode, BtnThumbL, BtnThumbR).
Testing ... (interrupt to exit)
Axes:  0:0  1:0  2:-32767  3:0  4:0  5:-32767  6:0  7:0
Buttons:  0:off  1:off  2:off  3:off  4:off  5:off  6:off
          7:off  8:off  9:off 10:off 11:off 12:off
```

From the output, we can see the different buttons and axes that have been detected. We can see that all the buttons are off because none of them are pressed. Each button has a specific number used to identify it. The value of each button is either on or off. If we press and hold down the cross button on a PlayStation controller, we get the following output:

```
Buttons:  0:on   1:off  2:off  3:off  4:off  5:off  6:off
          7:off  8:off  9:off 10:off 11:off 12:off
```

We can see that the button number `0` is on, which means the cross button is mapped to this button. If we now press the circle button, we will get the following output:

```
Buttons:  0:off  1:on   2:off  3:off  4:off  5:off  6:off
          7:off  8:off  9:off 10:off 11:off 12:off
```

The output indicates the circle button is mapped to button number `1`. If we continue this process, it will show that the triangle button is mapped to button number `2`, and the square button is mapped to button number `3`. We can make note of these mappings and use them in our application to map the button numbers to button labels. If you are using an Xbox controller, you can follow the same procedure for the A, B, X, and Y buttons.

Next, let's explore the axes values. When nothing is pressed, we get the following values:

```
Axes:  0:0  1:0  2:-32767  3:0  4:0  5:-32767  6:0  7:0
```

From the documentation, we can see that when the stick is in the center position, the value of the axis is `0`. When the stick is pushed to the furthest direction along a specific axis, the value is `32767`, and when put in the opposite direction, it becomes `-32767`. Other positions of the stick will be represented between these values, depending on how far the stick is from the center. We can see from the output that two axes have a value of `-32767`, even though we are not moving the stick. The reason for this is that these two axes are mapped to the trigger buttons on the controller, which have hardware that can detect to what extent the button has been pressed, unlike the other buttons on the joystick. We don't need to use the triggers for our robot application, so we can ignore them. If we push the right stick to the most forward position, we get the following output:

```
Axes:  0:0  1:0  2:-32767  3:0  4:-32767  5:-32767  6:0  7:0
```

We can see that the *y*-axis for the right stick is mapped to axis number `4`, which has the value `-32767`. This means the forward position in the *y*-axis is mapped to the value `-32767`. Let us now move the right stick to the furthest position backward and see the results:

```
Axes:  0:0  1:0  2:-32767  3:0  4:32767  5:-32767  6:0  7:0
```

The same axis now has the value `32767`. The backward position is mapped to the value `32767`. By doing the same process for the *x*-axis on the right stick, we see it is mapped to axis number `3`. Similarly, we can discover that the left stick *x*-axis is mapped to axis number `0`, and the *y*-axis is mapped to axis number `1`.

The documentation has an excellent section called "Programming interface" that will give us everything we need to write our application. The general approach will be to open the joystick device file in the binary mode and read a fixed length of bytes from the device file. Each chunk of data we read is a single joystick event. The structure of the binary data we are reading is in the following format:

```
struct js_event {
    __u32 time;     /* event timestamp in milliseconds */
    __s16 value;    /* value */
    __u8 type;      /* event type */
    __u8 number;    /* axis/button number */
};
```

Python has a built-in module to read binary data in the C language `struct` format and convert it to its related Python value. At this stage, we just need to make note of the data types and their meanings. The first value is a time stamp that we won't need. Then, `value` will have the same button and axis values we saw using the `jstest` command. By inspecting `type`, we can tell whether the event is a button or an axis event. The documentation states that for the `type` variable, button events will have a value of `1` and axis events will have a value of `2`. Finally, `number` indicates which button or axis the event is for in the same way we saw for the `jstest` output. Now, we have all the required information to put together our implementation.

7.5.2 *Unpacking joystick events*

This script will have the core logic of reading joystick events from the Linux input subsystem and converting the event data to Python. We will import the `struct` module that is part of the Python standard library. This module provides the functionality to convert C structs data to Python values:

```
from struct import Struct
```

We then save the path to the joystick device file in a variable called `DEVICE`:

```
DEVICE = '/dev/input/js0'
```

We now define the function `main`, which first creates a `Struct` object and saves it in the `event_struct` variable. The Python documentation for the `struct` module shows how to map different C data types. The first value is a `u32`, which is an unsigned integer of the 32-bit length, and so it maps to `I`. The next value is a `__s16`, which is a 16-bit integer and thus maps to `h`. Finally, the last two values are both `u8`, which is an 8-bit integer and thus maps to `B`. This makes the `Struct` object created with the format `'I h B B.'` We now open the device file with the mode `'rb'` so that the file can be opened for reading in the binary mode. We then enter into a `while` loop that continuously reads event data from the device file. Next, we read the `event_struct.size` number of bytes from the file and save it into the `bytes` variable. This value is the exact size of the C `structs` data in bytes. By reading this exact size, we are reading a single joystick event in each loop. Next, we use the `unpack` method to convert the data in bytes to a set of Python values that are saved in `data`. Then, we save each part of the event data into individual variables. Finally, we print out `value` and `number` when a button press event is detected, which correlates to `type` being `1`:

```
def main():
    event_struct = Struct('I h B B')
    with open(DEVICE, 'rb') as js_device:
        while True:
            bytes = js_device.read(event_struct.size)
            data = event_struct.unpack(bytes)
            time, value, type, number = data
            if type == 1:
                print(f'value:{value} number:{number}')
```

The last line of the application calls the `main` function:

```
main()
```

The full script can be saved as `joystick_unpack.py` on the Pi and then executed.

Listing 7.3 `joystick_unpack.py`: Unpacking raw joystick events on Linux

```
#!/usr/bin/env python3
from struct import Struct

DEVICE = '/dev/input/js0'

def main():
    event_struct = Struct('I h B B')
    with open(DEVICE, 'rb') as js_device:
        while True:
            bytes = js_device.read(event_struct.size)
            data = event_struct.unpack(bytes)
            time, value, type, number = data
            if type == 1:
                print(f'value:{value} number:{number}')

main()
```

The output shows a session where the script was run and then the cross and circle button were pressed. The first two lines of output relate to the cross button that has `number:0` as its button number. We can see that the `value` for the button changes from `1` to `0` as it is pressed and then released. The last two lines show the same, except with `number:1`, which indicates the circle button being pressed and released:

```
$ joystick_unpack.py
value:1 number:0
value:0 number:0
value:1 number:1
value:0 number:1
```

7.5.3 Mapping joystick events

The next step is mapping the values in the events to more readable names for the buttons and axes, as well as the event types. This will make our code more readable and provide us with the ability to produce more readable output in the terminal. We will also create a dedicated function that will be called to handle joystick events as we receive them.

We will import the `namedtuple` object from the `collections` module, which is part of the Python standard library. This object provides a great way to convert a Python `tuple` object to a more readable `namedtuple`:

```
from collections import namedtuple
```

We save the values for button and axis type events in `TYPE_BUTTON` and `TYPE_AXIS`. We use a dictionary to map the name of each button in the `BUTTON` variable. The first version of BUTTON has the mapping for the PlayStation controller, while the second commented-out version has the mapping for the Xbox controller. You can use either one as needed. Then, we create a dictionary called AXIS to obtain the name of an axis for axis events:

```
TYPE_BUTTON = 1
TYPE_AXIS = 2
BUTTON = {0: 'cross', 1: 'circle', 2: 'triangle', 3: 'square'}
# BUTTON = {0: 'A', 1: 'B', 2: 'X', 3: 'Y'}
AXIS = {0: 'left_x', 1: 'left_y', 3: 'right_x', 4: 'right_y'}
```

A namedtuple called Event is created, and it will be used to save event data in a more readable data structure:

```
Event = namedtuple('Event', 'time, value, type, number')
```

The main function is largely the same as the previous script. The main difference is that an Event object is created for each new event, and then the handle_event function is called with this object:

```
def main():
    event_struct = Struct('I h B B')
    with open(DEVICE, 'rb') as js_device:
        while True:
            bytes = js_device.read(event_struct.size)
            event = Event(*event_struct.unpack(bytes))
            handle_event(event)
```

When the handle_event function encounters a button event, it will look up the name of the button using the BUTTON dictionary. We use the get method so that if a button we haven't defined is pressed, it doesn't cause an error; instead, the None value is returned. In this way, we can define the names of buttons we care most about. We then output to the terminal that a button event was encountered and provide the button name and dump of the event variable. When an axis event is detected, the name of the axis is retrieved, and similar details of the axis name and event variable are outputted:

```
def handle_event(event):
    if event.type == TYPE_BUTTON:
        name = BUTTON.get(event.number)
        print('button -', name, event)
    if event.type == TYPE_AXIS:
        name = AXIS.get(event.number)
        print('axis -', name, event)
```

The full script can be saved as joystick_map.py on the Pi and then executed.

Listing 7.4 joystick_map.py: Map joystick events to button and axis names

```
#!/usr/bin/env python3
from struct import Struct
from collections import namedtuple
```

```
DEVICE = '/dev/input/js0'
TYPE_BUTTON = 1
TYPE_AXIS = 2
BUTTON = {0: 'cross', 1: 'circle', 2: 'triangle', 3: 'square'}
# BUTTON = {0: 'A', 1: 'B', 2: 'X', 3: 'Y'}
AXIS = {0: 'left_x', 1: 'left_y', 3: 'right_x', 4: 'right_y'}

Event = namedtuple('Event', 'time, value, type, number')

def handle_event(event):
    if event.type == TYPE_BUTTON:
        name = BUTTON.get(event.number)
        print('button -', name, event)
    if event.type == TYPE_AXIS:
        name = AXIS.get(event.number)
        print('axis -', name, event)

def main():
    event_struct = Struct('I h B B')
    with open(DEVICE, 'rb') as js_device:
        while True:
            bytes = js_device.read(event_struct.size)
            event = Event(*event_struct.unpack(bytes))
            handle_event(event)

main()
```

The following session shows the output when the cross and circle buttons are pressed. Then, the right stick is moved along the *y*- and the *x*-axes. Finally, the left stick is moved again on both the axes:

```
$ joystick_map.py
button - cross Event(time=16675880, value=1, type=1, number=0)
button - cross Event(time=16676010, value=0, type=1, number=0)
button - circle Event(time=16676400, value=1, type=1, number=1)
button - circle Event(time=16676500, value=0, type=1, number=1)
axis - right_x Event(time=16677540, value=-1014, type=2, number=3)
axis - right_y Event(time=16677540, value=-1014, type=2, number=4)
axis - right_x Event(time=16677540, value=-2365, type=2, number=3)
axis - right_y Event(time=16677540, value=-2027, type=2, number=4)
axis - right_x Event(time=16677550, value=-3041, type=2, number=3)
axis - right_y Event(time=16677550, value=-2703, type=2, number=4)
axis - left_x Event(time=16681520, value=7769, type=2, number=0)
axis - left_y Event(time=16681520, value=-19932, type=2, number=1)
axis - left_x Event(time=16681520, value=5067, type=2, number=0)
axis - left_y Event(time=16681520, value=-12500, type=2, number=1)
axis - left_x Event(time=16681530, value=0, type=2, number=0)
axis - left_y Event(time=16681530, value=-2365, type=2, number=1)
```

7.5.4 Working with axis events

We can now delve more deeply into the axis events and calculate the direction and how far the stick was moved. For controlling the robot, we only care about the stick moving on the *y*-axis, so we will only focus on the events on that axis. The `MAX_VAL` variable can be used so that we can compare the stick position to the maximum possible value to calculate the percentage of movement:

```
MAX_VAL = 32767
```

The `handle_event` function has been changed to only focus on axis events. Once the name of the axis is obtained, it is checked to make sure that it is either `left_y` or `right_y`. In this way, only *y*-axis events are processed. The `direction` variable will keep track of whether the stick is being pushed forward or backward. This value is calculated based on whether `event.value` is negative or positive. The absolute value of `event.value` is taken and divided by `MAX_VAL` to calculate the fractional position of the stick away from center. This value is multiplied by a hundred and rounded to two decimal points to express it as a percentage. These three variables are then outputted to the terminal:

```
def handle_event(event):
    if event.type == TYPE_AXIS:
        name = AXIS.get(event.number)
        if name in ['left_y', 'right_y']:
            direction = 'backward' if event.value > 0 else 'forward'
            percent = round((abs(event.value) / MAX_VAL) * 100, 2)
            print(name, direction, percent)
```

The full script can be saved as `joystick_axis.py` on the Pi and then executed.

Listing 7.5 `joystick_axis.py`: Controlling direction and movement percentage

```
#!/usr/bin/env python3
from struct import Struct
from collections import namedtuple

DEVICE = '/dev/input/js0'
TYPE_AXIS = 2
AXIS = {0: 'left_x', 1: 'left_y', 3: 'right_x', 4: 'right_y'}
MAX_VAL = 32767

Event = namedtuple('Event', 'time, value, type, number')

def handle_event(event):
    if event.type == TYPE_AXIS:
        name = AXIS.get(event.number)
        if name in ['left_y', 'right_y']:
            direction = 'backward' if event.value > 0 else 'forward'
            percent = round((abs(event.value) / MAX_VAL) * 100, 2)
            print(name, direction, percent)

def main():
    event_struct = Struct('I h B B')
    with open(DEVICE, 'rb') as js_device:
        while True:
            bytes = js_device.read(event_struct.size)
```

```
            event = Event(*event_struct.unpack(bytes))
            handle_event(event)

main()
```

The following session shows the output generated during script execution when the right stick is moved forward to the fullest position and then back to center. The right stick is then moved backward a bit and back to center. Finally, the left stick is also moved forward and backward:

```
$ joystick_axis.py
right_y forward 13.4
right_y forward 40.21
right_y forward 79.38
right_y forward 100.0
right_y forward 95.88
right_y forward 35.05
right_y forward 0.0
right_y backward 18.56
right_y backward 48.45
right_y backward 22.68
right_y backward 9.28
right_y forward 0.0
left_y forward 29.9
left_y forward 56.7
left_y forward 83.51
left_y forward 100.0
left_y forward 97.94
left_y forward 86.6
left_y forward 28.87
left_y forward 0.0
left_y backward 16.5
left_y backward 28.86
left_y backward 41.24
left_y backward 19.59
left_y forward 0.0
```

7.6 Measuring the rate of joystick events

The sensors on the analog sticks are extremely sensitive and can detect hundreds of different positions. Such sensitivity generates a very high rate of joystick events per second. We want to be mindful of this, as it could flood our robot with requests to change the throttle on the motors even for very small changes in stick positions. In the `motor` module, we have included three levels of speed. We can apply a similar approach to solve our joystick problem by calculating three levels of joystick position for each direction and only requesting a robot movement when one of these levels has changed. Each of the levels will be directly correlated with a speed level.

Another common solution used with joysticks is creating a configurable joystick dead zone. This dead zone is how far the stick needs to be moved from the neutral position before the application will treat it as a movement.

We will write a script to measure and report how many axis events are generated per second so that we can quantify the problem. Then, we will enhance the script to calculate changes on three defined levels and the rate at which the levels change. With these measurements, we will have data to conclude whether this solution solves the problem. The script will focus only on the *y*-axis on a single stick to make implementation and measurement simpler. The application will need to meet the following requirements:

- A Python application should read 100 axis events and record the time taken.
- The event rate should be calculated as new events per second.
- The level rate should be calculated as new levels per second.
- Only axis movements on the right-stick *y*-axis should be recorded.

7.6.1 Calculating the event rate

The initial task at hand will be to calculate the number of new axis events per second. The script will build on the work from the script in the previous section. We will import the `sys` module so that we can exit the script when all measurements have been taken. The `time` module is imported to measure the time. The `SimpleNamespace` object is imported from the `types` module and will be used to keep track of our statistical data:

```
import sys
import time
from types import SimpleNamespace
```

The `main` function is very similar to what we have seen in the previous script, with some small additions. The `data` variable will keep track of the number of events encountered so far and the start time of the first event:

```
def main():
    data = SimpleNamespace(events=0, start=None)
    event_struct = Struct('I h B B')
    with open(DEVICE, 'rb') as js_device:
        while True:
            bytes = js_device.read(event_struct.size)
            event = Event(*event_struct.unpack(bytes))
            handle_event(event, data)
```

The `handle_event` will check for any *y*-axis events on the right stick. Each time it detects one, it will call the `update_counter` function so that it can update the counter statistics:

```
def handle_event(event, data):
    if event.type == TYPE_AXIS:
        name = AXIS.get(event.number)
        if name == 'right_y':
            update_counter(data)
```

The `update_counter` function increments the `data.events` counter variable to record the new event. The number of counted events is then outputted. If this is the first event encountered, then the starting time of the sample is saved in the `data.start` variable. If a hundred samples have been collected, then the `stop_counter` function is called to end measurement and report the results:

```
def update_counter(data):
    data.events += 1
    print('events:', data.events)
    if data.events == 1:
        data.start = time.perf_counter()
    if data.events == 100:
        stop_counter(data)
```

The `stop_counter` function first calculates how much time has elapsed while counting the new events and saving the results in `duration`. Then, the number of new events per second is calculated and saved in `event_rate`. Finally, the time taken, total number of events, and event rate are printed, and the script is exited:

```
def stop_counter(data):
    duration = time.perf_counter() - data.start
    event_rate = data.events / duration
    print('---------- STATS ----------')
    print(f'      time: {duration:0.3f}s')
    print(f'    events: {data.events}')
    print(f'event rate: {event_rate:0.1f}')
    sys.exit()
```

The full script can be saved as `joystick_stats.py` on the Pi and then executed.

Listing 7.6 `joystick_stats.py`: Collecting and reporting joystick statistics

```
#!/usr/bin/env python3
import sys
import time
from struct import Struct
from collections import namedtuple
from types import SimpleNamespace

DEVICE = '/dev/input/js0'
TYPE_AXIS = 2
AXIS = {0: 'left_x', 1: 'left_y', 3: 'right_x', 4: 'right_y'}
MAX_VAL = 32767

Event = namedtuple('Event', 'time, value, type, number')

def stop_counter(data):
    duration = time.perf_counter() - data.start
    event_rate = data.events / duration
    print('---------- STATS ----------')
    print(f'      time: {duration:0.3f}s')
    print(f'    events: {data.events}')
    print(f'event rate: {event_rate:0.1f}')
    sys.exit()

def update_counter(data):
    data.events += 1
    print('events:', data.events)
    if data.events == 1:
        data.start = time.perf_counter()
```

```
    if data.events == 100:
        stop_counter(data)

def handle_event(event, data):
    if event.type == TYPE_AXIS:
        name = AXIS.get(event.number)
        if name == 'right_y':
            update_counter(data)

def main():
    data = SimpleNamespace(events=0, start=None)
    event_struct = Struct('I h B B')
    with open(DEVICE, 'rb') as js_device:
        while True:
            bytes = js_device.read(event_struct.size)
            event = Event(*event_struct.unpack(bytes))
            handle_event(event, data)

main()
```

When you execute the script, take the right stick and continually push it forward and backward until 100 events have been detected. Then, the script will exit and print the measurement results. The session shows the results of axis measurements that were taken. As expected, the number of events per second is quite high for our purposes and could pose a challenge for our robot motor server to keep up with if we send a throttle request for every stick movement:

```
$ joystick_stats.py
events: 1
events: 2
events: 3
...
events: 98
events: 99
events: 100
---------- STATS ----------
      time: 0.722s
    events: 100
event rate: 138.5
```

7.6.2 Calculating the level rate

We can now enhance our script to calculate the three different levels of the stick position for each direction of movement along the *y*-axis. Then, we can calculate the number of new levels per second.

The `data` variable has some additional attributes to keep track of the number of new levels and the last level that was encountered. The rest of the function is unchanged:

```
def main():
    data = SimpleNamespace(events=0, levels=0, last_level=0, start=None)
    event_struct = Struct('I h B B')
    with open(DEVICE, 'rb') as js_device:
        while True:
```

```
            bytes = js_device.read(event_struct.size)
            event = Event(*event_struct.unpack(bytes))
            handle_event(event, data)
```

The `handle_event` function will calculate the level by taking the distance the stick has moved from the center and dividing it by three. This value is then saved in the `level` variable and given to the `update_counter` function when called:

```
def handle_event(event, data):
    if event.type == TYPE_AXIS:
        name = AXIS.get(event.number)
        if name == 'right_y':
            level = round((event.value / MAX_VAL) * 3)
            update_counter(data, level)
```

The `update_counter` function will now also increment `data.levels` every time a new level is encountered. On each call, the event and level counts are printed. The rest of the logic in the function remains the same:

```
def update_counter(data, level):
    data.events += 1
    if data.last_level != level:
        data.last_level = level
        data.levels += 1
    print('events:', data.events, 'level:', level)
    if data.events == 1:
        data.start = time.perf_counter()
    if data.events == 100:
        stop_counter(data)
```

The `stop_counter` function will now also calculate the level rate and output the rate and total number of new levels encountered:

```
def stop_counter(data):
    duration = time.perf_counter() - data.start
    event_rate = data.events / duration
    level_rate = data.levels / duration
    print('---------- STATS ----------')
    print(f'      time: {duration:0.3f}s')
    print(f'    events: {data.events}')
    print(f'event rate: {event_rate:0.1f}')
    print(f'    levels: {data.levels}')
    print(f'level rate: {level_rate:0.1f}')
    sys.exit()
```

The full script can be saved as `joystick_levels.py` on the Pi and then executed.

Listing 7.7 `joystick_levels.py`: Applying levels to joystick movements

```
#!/usr/bin/env python3
#!/usr/bin/env python3
import sys
import time
from struct import Struct
```

```
from collections import namedtuple
from types import SimpleNamespace

DEVICE = '/dev/input/js0'
TYPE_AXIS = 2
AXIS = {0: 'left_x', 1: 'left_y', 3: 'right_x', 4: 'right_y'}
MAX_VAL = 32767

Event = namedtuple('Event', 'time, value, type, number')

def stop_counter(data):
    duration = time.perf_counter() - data.start
    event_rate = data.events / duration
    level_rate = data.levels / duration
    print('---------- STATS ----------')
    print(f'       time: {duration:0.3f}s')
    print(f'     events: {data.events}')
    print(f'event rate: {event_rate:0.1f}')
    print(f'     levels: {data.levels}')
    print(f'level rate: {level_rate:0.1f}')
    sys.exit()

def update_counter(data, level):
    data.events += 1
    if data.last_level != level:
        data.last_level = level
        data.levels += 1
    print('events:', data.events, 'level:', level)
    if data.events == 1:
        data.start = time.perf_counter()
    if data.events == 100:
        stop_counter(data)

def handle_event(event, data):
    if event.type == TYPE_AXIS:
        name = AXIS.get(event.number)
        if name == 'right_y':
            level = round((event.value / MAX_VAL) * 3)
            update_counter(data, level)

def main():
    data = SimpleNamespace(events=0, levels=0, last_level=0, start=None)
    event_struct = Struct('I h B B')
    with open(DEVICE, 'rb') as js_device:
        while True:
            bytes = js_device.read(event_struct.size)
            event = Event(*event_struct.unpack(bytes))
            handle_event(event, data)

main()
```

We can now execute the new script and move the right stick again until 100 events have been detected. The session shows the results of event measurements that were taken again with level measurements being calculated this time. We can see from the results that the level rate of 10.3 is 14 times lower than the event rate of 146.6. By only sending robot movement requests when levels have changed instead of on every event, we can make very significant improvements on the request load on our robot motor server. With this performance enhancement, we still support all the three motor speed levels and will be able to provide a responsive experience for using the joystick across the network to move the robot:

```
$ joystick_levels.py
events: 1 level: 0
events: 2 level: 0
events: 3 level: 0
...
events: 98 level: -3
events: 99 level: -3
events: 100 level: -3
---------- STATS ----------
      time: 0.682s
    events: 100
event rate: 146.6
    levels: 7
level rate: 10.3
```

7.7 Moving robots with joysticks

We now have what we need to connect our joystick to our robot. The task at hand is to create an application that will meet the following requirements:

- We should create a Python application that moves the right motor forward and backward when the right stick moves forward and backward.
- The left motor should be moved based on left-stick movements in the same fashion.
- The joystick application should send the movement requests over HTTP to the robot server running the `robows` server.

7.7.1 Creating the joystick client

All the logic to send movement requests to the `robows` server is taken from `client_persist.py` in listing 5.6. The `json` and `http.client` modules are imported to make the needed HTTP requests. The `Struct` and `namedtuple` objects are imported to help in reading the joystick event data:

```
import json
from http.client import HTTPConnection
from struct import Struct
from collections import namedtuple
```

The `MOTOR_AXIS` dictionary is used to map the joystick axis to its associated motor on the robot. The left stick *y*-axis maps to the left motor, and the right stick *y*-axis maps to the right motor:

```
MOTOR_AXIS = dict(left_y='L', right_y='R')
```

The main function is very similar to the ones we saw previously in this chapter. An HTTP connection to the robot server is made and saved in HTTPConnection. Then, a dictionary called data is created to keep track of the position levels on the left and right sticks *y*-axis. The rest of the function is unchanged:

```
def main():
    conn = HTTPConnection('robopi:8888')
    data = dict(left_y=0, right_y=0)
    event_struct = Struct('I h B B')
    with open(DEVICE, 'rb') as js_device:
        while True:
            bytes = js_device.read(event_struct.size)
            event = Event(*event_struct.unpack(bytes))
            handle_event(event, data, conn)
```

The handle_event function will be called each time a new event is detected. It will check whether the event is an axis event on the *y*-axis of either the stick or right stick. If so, it will then calculate the level of the stick from the center position. The previous level is compared, and if the level has changed, it will then prepare to make a change to the throttle on the robot motor. The new level is saved in data. The name of the motor to move is looked up using MOTOR_AXIS. The value of factor is calculated based on whether the level is positive or negative. This will dictate whether the motor is turned in the forward or backward direction. The speed of the motor movement will be based on the absolute value of the level. These arguments are saved in a dictionary called args. The call_robot function is then called so that a request to call set_throttle on the robot server is made:

```
def handle_event(event, data, conn):
    if event.type == TYPE_AXIS:
        name = AXIS.get(event.number)
        if name in ['left_y', 'right_y']:
            level = round((event.value / MAX_VAL) * 3)
            if data[name] != level:
                print('level change:', name, level)
                data[name] = level
                motor = MOTOR_AXIS[name]
                factor = 1 if level <= 0 else -1
                args = dict(name=motor, speed=abs(level), factor=factor)
                call_robot(conn, 'set_throttle', **args)
```

The call_api and call_robot functions are as presented in chapter 5. They will send an HTTP request to the robot server to request changes in the motor throttle:

```
def call_api(conn, url, data):
    body = json.dumps(data).encode()
    conn.request('POST', url, body)
    with conn.getresponse() as resp:
        resp.read()

def call_robot(conn, func, **args):
    return call_api(conn, '/' + func, args)
```

The full script can be saved as `joystick_remote.py` on the Pi and then executed.

Listing 7.8 `joystick_remote.py`: Remotely controlling the robot motors

```
#!/usr/bin/env python3
import json
from http.client import HTTPConnection
from struct import Struct
from collections import namedtuple

DEVICE = '/dev/input/js0'
TYPE_AXIS = 2
AXIS = {0: 'left_x', 1: 'left_y', 3: 'right_x', 4: 'right_y'}
MOTOR_AXIS = dict(left_y='L', right_y='R')
MAX_VAL = 32767

Event = namedtuple('Event', 'time, value, type, number')

def call_api(conn, url, data):
    body = json.dumps(data).encode()
    conn.request('POST', url, body)
    with conn.getresponse() as resp:
        resp.read()

def call_robot(conn, func, **args):
    return call_api(conn, '/' + func, args)

def handle_event(event, data, conn):
    if event.type == TYPE_AXIS:
        name = AXIS.get(event.number)
        if name in ['left_y', 'right_y']:
            level = round((event.value / MAX_VAL) * 3)
            if data[name] != level:
                print('level change:', name, level)
                data[name] = level
                motor = MOTOR_AXIS[name]
                factor = 1 if level <= 0 else -1
                args = dict(name=motor, speed=abs(level), factor=factor)
                call_robot(conn, 'set_throttle', **args)

def main():
    conn = HTTPConnection('robopi:8888')
    data = dict(left_y=0, right_y=0)
    event_struct = Struct('I h B B')
    with open(DEVICE, 'rb') as js_device:
        while True:
            bytes = js_device.read(event_struct.size)
            event = Event(*event_struct.unpack(bytes))
            handle_event(event, data, conn)

main()
```

As was done in chapter 5, make sure to keep the `robows` server running in one terminal and then run this script in another. In the following session, we can see the output from the script. The right stick was pushed all the way forward and then brought back to center to make the right motor move forward. Then, the right stick was pulled backward and returned to center to make the right motor turn backward. Finally, both the right and left sticks were pushed forward together to make both motors drive forward, and then both were returned to center to make the robot come to a full stop:

```
$ joystick_remote.py
level change: right_y -1
level change: right_y -2
level change: right_y -3
level change: right_y -2
level change: right_y -1
level change: right_y 0
level change: right_y 1
level change: right_y 2
level change: right_y 3
level change: right_y 2
level change: right_y 1
level change: right_y 0
level change: left_y -1
level change: left_y -2
level change: left_y -3
level change: right_y -1
level change: right_y -2
level change: right_y -3
level change: right_y 0
level change: left_y -1
level change: left_y 1
level change: left_y 0
```

Robots in the real world: Remote forklift operators

A number of robotics startup companies have begun providing services for vehicles such as forklifts to be operated remotely. The remote driver can see the area around the vehicle using multiple cameras and drive the forklift using a joystick controller. This opens the potential to match drivers from many remote locations to different sites, which has become an important technology for addressing shortages in heavy vehicle drivers. The BBC article (https://www.bbc.com/news/business-54431056) on the subject mentions many of the opportunities this approach offers, as well as the risks.

Safety features, such as having microphones on the vehicles so that remote drivers can hear if anyone around the vehicle is calling for them to stop the vehicle, are essential. Security measures are also put in place at the software and network levels to prevent unauthorized access to the vehicles by malicious parties.

Summary

- Joysticks offer superior control over keyboards and mice when used to control the motors of a robot.
- The joystick hardware can be connected to the Raspberry Pi using either a wired USB or a wireless Bluetooth connection.
- The `struct` Python module can be used to directly read and parse joystick events from the Linux input subsystem.
- Pygame is a very popular Python module used for writing video games.
- Linux provides very good support for joystick devices, and their event data can be read directly by Python applications.
- The sensors on the analog sticks of the controller are extremely sensitive and can detect hundreds of different positions.
- To remotely control the robot motors, the joystick application must send movement requests over HTTP to the robot server running the `robows` server.

8 Keyboard-controlled camera

This chapter covers

- Capturing images and streaming live videos from the camera
- Drawing shapes and writing text using the OpenCV library
- Moving servo motors to specific angles and positions
- Using keyboard events in OpenCV to make servo movements

In this chapter, we will build a robot with a camera attached to two servo motors. One servo will allow us to pan the camera, and the other will apply a tilt motion. In this way, we can point the camera in many different directions. In addition, we will detect and use keyboard events to move the motors in different directions and capture and save photos from the live camera stream.

We have explored the use of touch screens on mobile devices and joysticks as human interaction devices. Now we will use a keyboard to control our robot movements and camera. Keyboards are one of the oldest and most established input devices. Compared to joysticks, they come with every desktop computer and laptop and have excellent built-in support in most software, not requiring much additional effort. This makes them an excellent alternative to joysticks when we either don't want to add additional hardware requirements or just want to simplify our device handling.

8.1 Hardware stack

Figure 8.1 shows the hardware stack, with the specific components used in this chapter highlighted. The camera can be directly connected to the Raspberry Pi using the camera connector. The camera used in this chapter is the Raspberry Pi Camera Module 2.

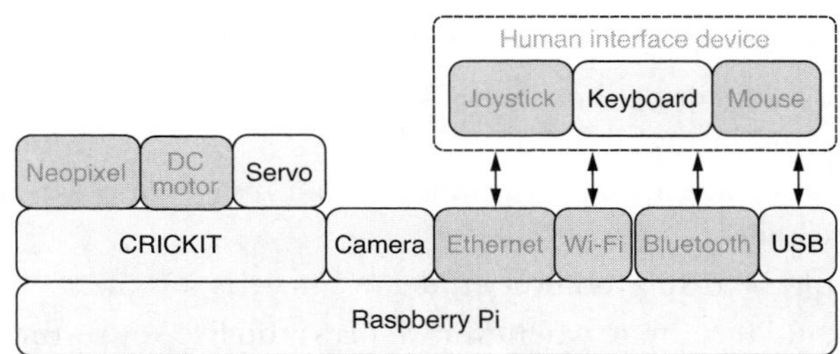

Figure 8.1 Hardware stack: the camera and servo will be controlled using the keyboard.

The servo motors are part of a Pan-Tilt Kit that comes fully assembled. The camera will be mounted to this kit. For further details on the robot assembly, check the robot assembly guide in appendix C. It shows how to assemble the robot used in this chapter. Also, make sure to check the hardware purchasing guide in appendix A before buying the hardware needed in this chapter. Any keyboard can be used for the robot; whether it is a USB or Bluetooth keyboard, there are no special requirements.

8.2 Software stack

Details of the specific software used in this chapter are described in figure 8.2. The `draw` and `snapshot` applications in this chapter will use the OpenCV and NumPy libraries to capture images from the camera and draw shapes and write text on

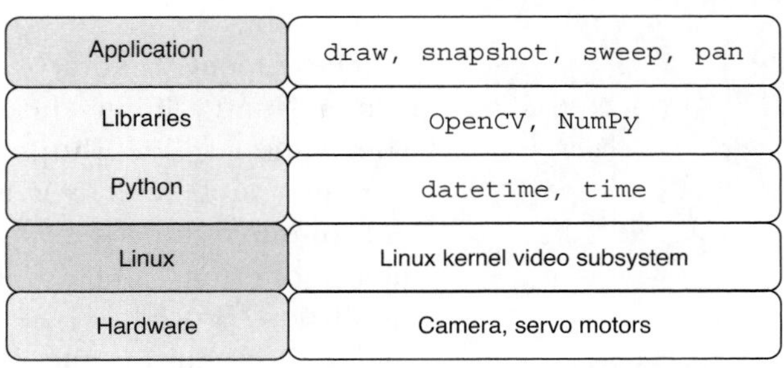

Figure 8.2 Software stack: the OpenCV library will capture images from the camera.

images. We will also learn how to read keyboard events using the OpenCV library and perform actions such as taking photos when specific keyboard events are detected. The OpenCV library will be used to interact with the Linux kernel video subsystem to capture images from the camera hardware. The `sweep` and `pan` applications come later in the chapter and will let us control the servo motor hardware, as well as take photos with the camera.

8.3 Capturing images using OpenCV

OpenCV is a very popular and powerful computer vision library. We will use it to interact with the camera to show a live video stream and save snapshots. In later chapters, we will expand our usage to perform face detection and QR code detection and decoding. We need to create an application to meet the following requirements:

- A Python application that uses the OpenCV library to show a live video feed from the camera should be created.
- It should be possible to save a snapshot image with a time stamp whenever the user presses the spacebar.
- A text message should be displayed every time an image is saved.
- The application should terminate whenever we press the Esc key or the letter Q.

We are creating this application to have some fundamental pieces in place that we can use to build more complex applications. Interacting with live video from the camera will give us exposure to the camera frame rates and prepare for face and QR code detection from a live video feed. Doing all these operations using OpenCV is strategic, as it is the library of choice for computer vision in Python.

8.3.1 Exploring the OpenCV library

To work with the camera in OpenCV, we will first have to enable legacy camera support on the Raspberry Pi. Use the `raspi-config` in a terminal and enable the Legacy Camera option in the Interface Options menu. Figure 8.3 shows a photo of the Raspberry Pi Camera placed in an Adafruit camera case to protect it and make it easier to attach to the robot.

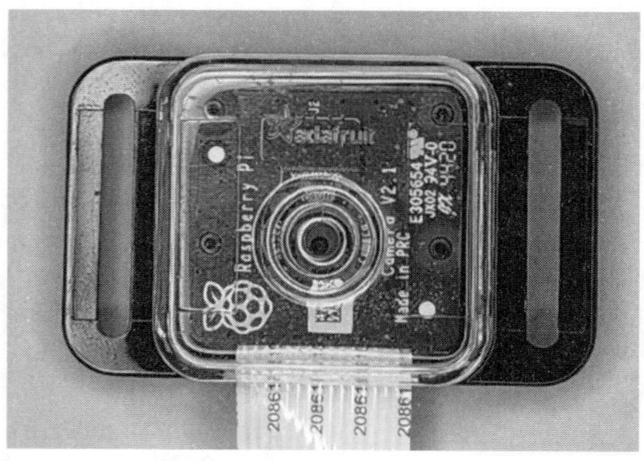

Figure 8.3 Raspberry Pi camera: the camera is enclosed in an Adafruit camera case.

The OpenCV library uses the Python NumPy library heavily. The NumPy user guide (https://numpy.org/doc/stable/user/) is an excellent resource for NumPy management offering details on the installation procedure. As per the guide, it requires the `libatlas-base-dev` package to be installed before NumPy. OpenCV requires the `libgtk2.0-dev` package to support its graphical interface features. Run the following command to install these two packages:

```
$ sudo apt install libatlas-base-dev libgtk2.0-dev
```

We can now install OpenCV with the following command:

```
$ ~/pyenv/bin/pip install opencv-python
```

The same command also installs NumPy automatically, as it is an adjunct of OpenCV. OpenCV has a large code base, and its installation on the Raspberry Pi can take up to 70 min to complete; keep this in mind when you perform the installation.

We can now dive right into a REPL session and start interacting with OpenCV and the camera. We start our adventure by first importing `cv2` and calling `VideoCapture` with `0` as an argument to open the default camera. We save the `VideoCapture` object in a variable called `cap`:

```
>>> import cv2
>>> cap = cv2.VideoCapture(0)
```

We can check whether the camera was initialized correctly and is open by calling the `isOpened` method. It will return `True` if the camera was initialized correctly:

```
>>> cap.isOpened()
True
```

We can also inquire about different properties relating to the video capture device. The following session shows that the size of images to be captured is set at the width `640` and height `480`. We can also check the frame rate of the video, which is set at 30 frames per second in our case:

```
>>> cap.get(cv2.CAP_PROP_FRAME_WIDTH)
640.0
>>> cap.get(cv2.CAP_PROP_FRAME_HEIGHT)
480.0
>>> cap.get(cv2.CAP_PROP_FPS)
30.0
```

Next, we call the `read` method to grab and decode the next video frame. It will return two values. The first is a Boolean value indicating whether the call was successful. The second value is the image itself. By inspecting the value of `ret`, we can see that the call was successful. If we check what data type `frame` is, we see it reported as a `numpy.ndarray`. This data structure is at the heart of NumPy and provides an n-dimensional array of data types that perform very efficiently:

```
>>> ret, frame = cap.read()
>>> ret
True
>>> type(frame)
<class 'numpy.ndarray'>
```

Whenever we work with images in OpenCV, we will be using `ndarray` objects. There are some useful attributes we can inspect to get more details about the image we have just captured. The `shape` attribute is showing us that as expected, our image has a height of `480` and a width of `640`. The last part showing `3` indicates that it is a color image and has the three color components: red, green and blue. The `dtype` indicates the data type of each item in the array. The `uint8` data type shows that each value is an 8-bit integer, which relates to each color component being a value ranging from 0 to 255:

```
>>> frame.shape
(480, 640, 3)
>>> frame.dtype
dtype('uint8')
```

We can now save our image data to disk by calling the `imwrite` method with our filename. This method will use the file extension to encode the image data in the expected image format. Next, we will find an image called `photo.jpg` in our current directory that is a snapshot taken from the camera:

```
>>> cv2.imwrite('photo.jpg', frame)
True
```

Once we are done working with the camera, it is always a good idea to close the video stream in a smooth fashion by calling the `release` method. We can see that the video stream state is closed when we call `isOpened` after closing the capture device:

```
>>> cap.release()
>>> cap.isOpened()
False
```

This gave us some good exposure to both OpenCV and NumPy. We were able to get familiar with the core objects and operations required for capturing images using OpenCV.

8.3.2 Drawing shapes and displaying text with OpenCV

When performing computer vision activities such as facial detection, it is very useful to be able to draw shapes, such as rectangles, over the exact parts of an image that has been detected. Placing text on an image is another common drawing operation to display a message. We will write a script to call the main drawing functions in OpenCV and demonstrate how they are used. First, we import the `string` module from the Python standard library that we will use to display the lowercase letters. Then we import the `cv2` and `numpy` modules:

```
import string
import numpy as np
import cv2
```

Next, we define three colors as constants to be used for drawing our shapes. As in most systems, colors are represented by their red, green, and blue elements. Most systems use an RGB encoding, and it is important to keep in mind that the default color encoding in OpenCV is BGR:

```
BLUE = (255, 0, 0)
GREEN = (0, 255, 0)
RED = (0, 0, 255)
```

To create a new image, we call `np.zeros`, which will create a new `ndarray` object filled with zero values. When all the color components of an image have a zero value, the color will be black. The `shape` of the array is the same as the one we used with the captured image from the camera. This array is a black-colored BGR image with a width `640` and height `480`:

```
img = np.zeros(shape=(480, 640, 3), dtype=np.uint8)
```

We can now start drawing shapes on the image. The next line will draw a red circle with a radius of 100 px centered at the (x, y) coordinates `(200, 110)`:

```
cv2.circle(img, center=(200, 110), radius=100, color=RED)
```

Next, we draw a green line from the top-left part of the image at coordinates `(0, 0)` to the center of the circle at `(200, 110)`:

```
cv2.line(img, pt1=(0, 0), pt2=(200, 110), color=GREEN)
```

We then call `rectangle` to draw a blue box under the circle with one corner at `(50, 250)` and the other at `(350, 350)`. This creates a box with a width `300` and height `100`:

```
cv2.rectangle(img, pt1=(50, 250), pt2=(350, 350), color=BLUE)
```

The last thing we will place on the image is some text by calling the `putText` function. We will display the alphabet in lowercase letters using the `FONT_HERSHEY_SIMPLEX` font at the position `(10, 380)` with a normal scale and the color red:

```
text = string.ascii_lowercase
font = cv2.FONT_HERSHEY_SIMPLEX
pos = (10, 380)
cv2.putText(img, text, org=pos, fontFace=font, fontScale=1, color=RED)
```

The last two lines of the script will display the image and wait for a key to be pressed before exiting the application:

```
cv2.imshow('preview', img)
cv2.waitKey()
```

The full script can be saved as `draw.py` on the Pi and then executed.

Listing 8.1 `draw.py`: Drawing shapes of different sizes and positions using OpenCV

```
#!/usr/bin/env python3
import string
import numpy as np
import cv2
```

```
BLUE = (255, 0, 0)
GREEN = (0, 255, 0)
RED = (0, 0, 255)

img = np.zeros(shape=(480, 640, 3), dtype=np.uint8)

cv2.circle(img, center=(200, 110), radius=100, color=RED)
cv2.line(img, pt1=(0, 0), pt2=(200, 110), color=GREEN)
cv2.rectangle(img, pt1=(50, 250), pt2=(350, 350), color=BLUE)

text = string.ascii_lowercase
font = cv2.FONT_HERSHEY_SIMPLEX
pos = (10, 380)
cv2.putText(img, text, org=pos, fontFace=font, fontScale=1, color=RED)

cv2.imshow('preview', img)
cv2.waitKey()
```

This script will create a window, so it needs to be run in an environment that supports creating graphical windows. There are three options supported by OpenCV:

- Run the script directly on the Raspberry Pi in a desktop environment with a connected keyboard, mouse, and monitor.
- Run the script remotely over VNC using a VNC viewer.
- Run the script remotely with SSH X11 forwarding using `ssh robo@robopi -X`.

It is very handy to have multiple ways to run these graphical scripts, as you can choose whichever approach best suits the hardware and software you have on hand. Figure 8.4 shows what the window and image generated by the script will look like.

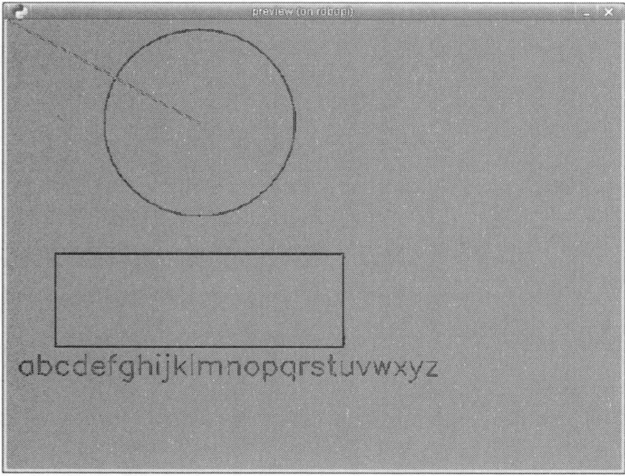

Figure 8.4 Drawing shapes: OpenCV supports drawing shapes and text on images.

Now that we have the fundamentals of drawing shapes using OpenCV out of the way, we can move on to the next task at hand. In the next sections, we will use some of these functions to help us display text in the application as we take snapshots.

8.3.3 Taking snapshots with OpenCV

We have now done enough exploration to create our camera application that will let us take photo snapshots from the live camera video feed. We import `datetime` to help us generate filenames with time stamps. We also import the OpenCV module:

```
from datetime import datetime
import cv2
```

For readability, we save the key code for the escape key in a constant called `ESC_KEY`. We also save the color code for blue and save the font we will use in `FONT`. When we display text messages in the application, we will position them in the top left corner using `TEXT_POS`. We will be capturing and displaying images from the camera at the default frame rate of 30 frames per second. We want to show our text messages for a specific number of frames. We save this setting in a constant called `MSG_FRAME_COUNT`. It is set at 10 frames, so it will show the messages for one-third of a second:

```
ESC_KEY = 27
BLUE = (255, 0, 0)
FONT = cv2.FONT_HERSHEY_SIMPLEX
TEXT_POS = (10, 30)
MSG_FRAME_COUNT = 10
```

The function `main` is at the heart of our application. It opens the default camera as a `VideoCapture` object in the variable `cap`. We then use the `assert` statement to make sure the camera was initialized correctly; otherwise, we exit with a `'Cannot open camera'` message. The `assert` statement is a great way to ensure certain conditions are met at different points in your program:

```
def main():
    cap = cv2.VideoCapture(0)
    assert cap.isOpened(), 'Cannot open camera'
```

In the same function, we then initialize the `messages` variables as an empty list that will store any text messages to be displayed in the application. Next, we enter our main event loop that calls `waitKey` during each loop. In the previous section, we called it with no arguments, so it would wait indefinitely until a key is pressed. Here, we call it as `waitKey(1)`, which will make it wait for 1 ms and return. If a key is pressed, it will return the code of the pressed key; otherwise, it will return the value `-1`. Calling `waitKey` also serves the important purpose of fetching and handling GUI events on the window. We save the returned value using the Python walrus operator into the variable `key`. We then check whether the key pressed is either the Esc key or the letter Q. The loop is then exited:

```
messages = []
while (key := cv2.waitKey(1)) not in [ord('q'), ESC_KEY]:
```

Once we enter the event loop, we capture an image from the camera and save it in `frame`. Like before, we use `assert` to make sure our call to `cap.read` was successful. If the space key was pressed, we call the `save_photo` function to save the image. We then call `set_message` to set the text message in the application to the value `'saving photo...'`. At the end of the loop, we call `show_image` to display the image. When the loop is exited, `cap.release` is called to close the video capture stream:

```
        ret, frame = cap.read()
        assert ret, 'Cannot read frame from camera'
        if key == ord(' '):
            save_photo(frame)
            set_message(messages, 'saving photo...')
        show_image(frame, messages)

    cap.release()
```

Next, we define the function `save_photo` that receives the image data in an argument called `frame`. We generate a time stamp using `datetime.now().isoformat()` and replace all the occurrences of `':'` with `'.'`. This is so that we can avoid putting the `:` character in the image filename. Some software does not play well with filenames that have colons in the filename. We then call `imwrite` to save the image data to the current directory with a time-stamped file name:

```
def save_photo(frame):
    stamp = datetime.now().isoformat().replace(':', '.')
    cv2.imwrite(f'photo_{stamp}.jpg', frame)
```

We now define `show_image` to display the image with the given text in `messages`. If `messages` has any text items, they are removed from the list by calling `pop` and are displayed on the image with `putText`. Then, the image is displayed by calling `imshow`:

```
def show_image(frame, messages):
    if messages:
        cv2.putText(frame, messages.pop(), TEXT_POS, FONT, 1, BLUE)
    cv2.imshow('preview', frame)
```

The function `set_message` is defined next. It will take the argument `text` and will set the contents of the `messages` list to be a list of the value `text` repeating 10 times as defined in the `MSG_FRAME_COUNT` variable. This will display that message text for 10 frames.

```
def set_message(messages, text):
    messages[:] = [text] * MSG_FRAME_COUNT
```

The full script can be saved as `snapshot.py` on the Pi and then executed.

Listing 8.2 `snapshot.py`: Taking a snapshot when the space key is pressed

```
#!/usr/bin/env python3
from datetime import datetime
import cv2

ESC_KEY = 27
BLUE = (255, 0, 0)
FONT = cv2.FONT_HERSHEY_SIMPLEX
```

```
TEXT_POS = (10, 30)
MSG_FRAME_COUNT = 10

def save_photo(frame):
    stamp = datetime.now().isoformat().replace(':', '.')
    cv2.imwrite(f'photo_{stamp}.jpg', frame)

def show_image(frame, messages):
    if messages:
        cv2.putText(frame, messages.pop(), TEXT_POS, FONT, 1, BLUE)
    cv2.imshow('preview', frame)

def set_message(messages, text):
    messages[:] = [text] * MSG_FRAME_COUNT

def main():
    cap = cv2.VideoCapture(0)
    assert cap.isOpened(), 'Cannot open camera'

    messages = []
    while (key := cv2.waitKey(1)) not in [ord('q'), ESC_KEY]:
        ret, frame = cap.read()
        assert ret, 'Cannot read frame from camera'
        if key == ord(' '):
            save_photo(frame)
            set_message(messages, 'saving photo...')
        show_image(frame, messages)

    cap.release()

if __name__ == "__main__":
    main()
```

When you run this application, it will show a window with images being captured and displayed at a rate of 30 frames per second, which will create a live video feed. You can wave your hands at the camera to test the latency and response of capturing and displaying the images in the application. Then, strike different poses and press the spacebar to take snapshots of each. You will find each photo saved in your current directory. You can then exit the application by pressing either the Esc key or the Q key on your keyboard. Figure 8.5 shows what the snapshot application will look like as a photo is being saved.

Figure 8.5 Taking a snapshot: a photo preview is shown in the snapshot application.

8.4 Moving a camera with servos

We will now combine OpenCV and the CRICKIT library by having OpenCV capture images from the camera and the CRICKIT library issue movement instructions to the servo motors. We need to create an application that meets the following requirements:

- Pressing the left and right keys will pan the camera with servos in that direction.
- Pressing the up and down keys will tilt the camera with the servos in that direction.
- Pressing the spacebar will save a snapshot image from the camera.
- Movement and snapshot actions should be displayed as text messages in the application.

Creating this application will let us combine the camera and motor libraries we have used so far to create an application that lets us control what the camera is pointing at using our keyboard and motors. We get a live preview of what the camera is seeing and then can use the keyboard to save snapshot images from the camera.

8.4.1 Exploring the servo motors with the CRICKIT library

Servo motors have a DC motor in them. They, however, also have a sensor built into the hardware that can detect their exact position. So unlike the DC motors that we have worked with in previous chapters, we can move servo motors to exact positions. This makes them ideal for applications such as robotic arms, where you want to move the arm to a specific place. The Pan-Tilt Kit that we will be using in this chapter comes with two servo motors. Figure 8.6 shows a photo of the Pan-Tilt Kit. With the camera attached to the kit, we will be able to move the camera in many different directions. The bottom servo will pan the camera left and right, while the top servo will tilt the camera up and down. In this way, we can use the motors to point the camera in many different directions. It can be difficult to move the servos in their full range of motion with the camera attached, as the camera has a ribbon that isn't always long enough for all the positions the kit will get in. For this reason, it is a good idea to remove the camera when you first experiment with the full range of motion the kit provides.

Figure 8.6 Adafruit Pan-Tilt Kit: the kit has two servo motors included.

We can use the read–evaluate–print loop (REPL) to dive right into the CRICKIT library. We first import the `crickit` module:

```
>>> from adafruit_crickit import crickit
```

The servo that will pan the kit is connected to the first servo connection and is accessed using `crickit.servo_1`. We move the servo by setting the angle we would like to position it in. By default, the CRICKIT library sets the lowest angle or position of a servo as `0` degrees and the highest as `180` degrees. We can set the servo position by setting a value to `angle`. Run the following in the REPL to move the servo to the lowest angle:

```
>>> crickit.servo_1.angle = 0
```

Now we can move the motor to the middle position by setting the value to `90`:

```
>>> crickit.servo_1.angle = 90
```

If we set the value to `180`, the servo will move to its highest position:

```
>>> crickit.servo_1.angle = 180
```

If we measure how far the physical servo actually moved, we will see it wasn't `180` degrees. The pan servo we are using has an actual range that goes from `0` to `142` degrees. The CRICKIT library has a feature where we can set the actual real-world value on a software level. Once we do this, the angle values we set in software will match the real-world angle values. We now move the servo back to the lowest position and then set the actuation range of the servo with the following lines:

```
>>> crickit.servo_1.angle = 0
>>> crickit.servo_1.actuation_range = 142
```

We can again move the servo to the lowest, middle, and highest angles:

```
>>> crickit.servo_1.angle = 0
>>> crickit.servo_1.angle = 142
>>> 142/2
71.0
>>> crickit.servo_1.angle = 71
```

If, however, we try to set a value beyond the actuation range, the library will raise a `ValueError` exception to stop us:

```
>>> crickit.servo_1.angle = 180
Traceback (most recent call last):
  File "<stdin>", line 1, in <module>
  File "/home/robo/pyenv/lib/python3.9/site-packages/adafruit_motor/
     servo.py", line 136, in angle
    raise ValueError("Angle out of range")
ValueError: Angle out of range
>>>
```

We can move the other servo as well to control the tilt position of the camera. The lowest tilt position is correlated with the highest angle value of `180`. Run the following line to set this position:

```
>>> crickit.servo_2.angle = 180
```

The highest tilt position that would point the camera upward is correlated with the angle value of `90`:

```
>>> crickit.servo_2.angle = 90
```

Unlike the pan, the tilt is more limited in range, and we won't need to set an actuation range for it. We now have enough info to put together a script to move our servos in different directions.

> **Going Deeper: Servo motors**
>
> The CRICKIT board supports many different servo motors. It can have up to four different servos connected at the same time. It is quite flexible in the types of servos you can connect, as it supports any 5V-powered servo motor. The Adafruit online store offers a wide range of servo motors (https://adafruit.com/category/232) in the servo section of their store. The two servos that come as part of the Pan-Tilt Kit used in the book are called micro servos. They are smaller and less powerful than other servos. The strength of a servo is specified by the servo torque rating. Some of the standard servo motors will have more than double the power of a micro servo. Depending on the weight a servo must carry or push, you might want to get a larger and more powerful servo.
>
> The Motor Selection Guide (http://mng.bz/YRRB) on the Adafruit site is a great resource for learning more about how servo motors work. It also provides good information on the variety of sizes, torque, and speed ratings of different servo motors.

8.4.2 Performing a pan-and-tilt sweep

Ultimately, we want to control the servo pan and tilt movements with our keyboard arrow keys. Something that will help us move in that direction would be to translate up- and down-movement requests to their associated angle changes in the tilt servo. Also, to do the same for the left and right movements on the pan servo, we will create a script that issues a bunch of movement commands in those four directions and then translates them to the needed servo angle changes. First, we import the `crickit` module to control the servos and the `time` module to pause between movements:

```
from adafruit_crickit import crickit
import time
```

We control how much of an angle change each movement will cause with `ANGLE_STEP`. Then we define `PAN` and `TILT` as data structures for each servo. Each one is a `dict` that uses `servo` to refer to its related servo motor, `min` to control the lowest allowed angle, and `max` to set the highest allowed motor. In the `dict`, the value of `start` will place the servo at that angle when the application starts, and `range` will be the value used to set the actuation range for that servo. `MOVE` maps each of the four movements to the related servo and the direction of movement for that servo:

```
ANGLE_STEP = 2
PAN = dict(servo=crickit.servo_1, min=30, max=110, start=70, range=142)
TILT = dict(servo=crickit.servo_2, min=90, max=180, start=180, range=180)
MOVE = dict(left=(PAN, -1), right=(PAN, 1), up=(TILT, -1), down=(TILT, 1))
```

When the script starts, we call `init_motors` to set the actuation range for each motor and position the motor in the specified starting angle:

```
def init_motors():
    for motor in [PAN, TILT]:
        motor['servo'].actuation_range = motor['range']
        motor['servo'].angle = motor['start']
```

We now define `move_motor` that will be called with one of the four accepted movements. It will then use the `MOVE` dictionary to look up the related servo and `factor` to set which way the angle will be changed. Next, we calculate `new_angle` that indicates what the new angle would be. We then check if the new angle is within the defined range of `min` and `max`. If the new angle is allowed, we apply it by setting the value of `angle` of the related servo:

```
def move_motor(direction):
    motor, factor = MOVE[direction]
    new_angle = motor['servo'].angle + (ANGLE_STEP * factor)
    if motor['min'] <= new_angle <= motor['max']:
        motor['servo'].angle = new_angle
```

The `main` function first calls `init_motors` to initialize each motor's actuation range and starting position. We then enter a loop for 20 iterations, making the camera move left and up in each. We print out details of our movements in each loop and then pause for 0.1 seconds before performing the next iteration of movements. The same style of loop is performed again, but for the right and down movements:

```
def main():
    init_motors()

    for i in range(20):
        print('moving left and up')
        move_motor('left')
        move_motor('up')
        time.sleep(0.1)

    for i in range(20):
        print('moving right and down')
        move_motor('right')
        move_motor('down')
        time.sleep(0.1)
```

The full script can be saved as `pan.py` on the Pi and then executed.

Listing 8.3 `pan.py`: Performing pan and tilt movements with the servo motors

```
#!/usr/bin/env python3
from adafruit_crickit import crickit
import time

ANGLE_STEP = 2
PAN = dict(servo=crickit.servo_1, min=30, max=110, start=70, range=142)
TILT = dict(servo=crickit.servo_2, min=90, max=180, start=180, range=180)
MOVE = dict(left=(PAN, -1), right=(PAN, 1), up=(TILT, -1), down=(TILT, 1))

def move_motor(direction):
    motor, factor = MOVE[direction]
    new_angle = motor['servo'].angle + (ANGLE_STEP * factor)
    if motor['min'] <= new_angle <= motor['max']:
        motor['servo'].angle = new_angle

def init_motors():
    for motor in [PAN, TILT]:
        motor['servo'].actuation_range = motor['range']
        motor['servo'].angle = motor['start']

def main():
    init_motors()

    for i in range(20):
        print('moving left and up')
        move_motor('left')
        move_motor('up')
        time.sleep(0.1)

    for i in range(20):
        print('moving right and down')
        move_motor('right')
        move_motor('down')
        time.sleep(0.1)

if __name__ == "__main__":
    main()
```

When we execute this script, it will run a movement demonstration of the servo movements. The first set of movements will move the camera left and up 20 times until it reaches the maximum position we have set for panning left. Then, we will move right and down 20 times back to our starting position. The following session shows the output generated by the script:

```
$ pan.py
moving left and up
moving left and up
moving left and up
...
```

```
moving right and down
moving right and down
moving right and down
```

> **Robots in the real world: Robotic arms**
>
> Servo motors are at the heart of robotic arms. The arms are made up of multiple joints, which give the arm full freedom of movement. The servos in this chapter gave the camera the ability to pan and tilt in any direction. In the case of robotic arms, four servos at the different joints would be enough to give the arm a full range of motion.
>
> The article (http://mng.bz/G99v) on robotic arms by Intel covers some of the many benefits of using robotic arms in different industries, ranging from manufacturing to agriculture. Older robotic arms were limited in their applications because of the lack of computer vision in software. This means the arms could only pick up items placed in exact locations and in a specific orientation. Combining robotics with powerful computer vision software gave the robots the ability to detect the location of objects and adjust to different orientations. The applications of robotics and computer vision go hand in hand to make more versatile robotics solutions.

8.4.3 Controlling servos and the camera together

We can now pull all our work together and create the final application that will let us move the camera around using our keyboard and take snapshots on command. We start by importing the modules we need. We use `datetime` for time stamps like before, `cv2` to work with the camera and keyboard, and the `crickit` module to control the servo motors. We will be able to reuse the functionality we have written in this chapter by importing the needed functions from our `snapshot` and `pan` modules:

```
from datetime import datetime
import cv2
from adafruit_crickit import crickit
from snapshot import save_photo, show_image, set_message
from pan import move_motor, init_motors
```

The `ESC_KEY` value is the key code for the Esc key, as we have seen before. The `ARROW_KEYS` dictionary will be used to map the key codes for the arrow keys to their related key names. The names of the keys also directly pair up with our four supported movement actions:

```
ESC_KEY = 27
ARROW_KEYS = {81: 'left', 82: 'up', 83: 'right', 84: 'down'}
```

The `handle_key` function will handle any key press events that occur. If the key pressed is the spacebar, the `save_photo` function will be called so as to save a snapshot. If one of the arrow keys is pressed, the `move_motor` function will be called with the associated key name that was pressed. After handling each of these key events, the `set_message` function will be called to update the text message displayed in the application:

```
def handle_key(key, frame, messages):
    if key == ord(' '):
        save_photo(frame)
        set_message(messages, 'saving photo...')
    elif key in ARROW_KEYS.keys():
        move_motor(ARROW_KEYS[key])
        set_message(messages, f'moving {ARROW_KEYS[key]}...')
```

Finally, the `main` function will first call `init_motors` to initialize the servo motors. Then, the video capture device will be created, and the `messages` list will be initialized. At this point, we enter our main event loop, which loops until the Esc or Q key is pressed. In each loop, a frame will be captured from the camera, and `handle_key` will be called to process any keyboard events. The last part of the loop is to call `show_image` to display the latest captured image and any text messages. When this loop is exited, `cap.release` is called to release the video capture device:

```
def main():
    init_motors()
    cap = cv2.VideoCapture(0)
    assert cap.isOpened(), 'Cannot open camera'

    messages = []
    while (key := cv2.waitKey(1)) not in [ord('q'), ESC_KEY]:
        ret, frame = cap.read()
        assert ret, 'Cannot read frame from camera'
        handle_key(key, frame, messages)
        show_image(frame, messages)

    cap.release()
```

The full script can be saved as `servocam.py` on the Pi and then executed.

Listing 8.4 `servocam.py`: Controlling camera position with a keyboard

```
#!/usr/bin/env python3
from datetime import datetime
import cv2
from adafruit_crickit import crickit
from snapshot import save_photo, show_image, set_message
from pan import move_motor, init_motors

ESC_KEY = 27
ARROW_KEYS = {81: 'left', 82: 'up', 83: 'right', 84: 'down'}

def handle_key(key, frame, messages):
    if key == ord(' '):
        save_photo(frame)
        set_message(messages, 'saving photo...')
    elif key in ARROW_KEYS.keys():
        move_motor(ARROW_KEYS[key])
        set_message(messages, f'moving {ARROW_KEYS[key]}...')
```

```
def main():
    init_motors()
    cap = cv2.VideoCapture(0)
    assert cap.isOpened(), 'Cannot open camera'

    messages = []
    while (key := cv2.waitKey(1)) not in [ord('q'), ESC_KEY]:
        ret, frame = cap.read()
        assert ret, 'Cannot read frame from camera'
        handle_key(key, frame, messages)
        show_image(frame, messages)

    cap.release()

main()
```

When we run this script, the pan and tilt servos will move to their starting positions. You can now press the left and right arrow keys to pan the camera. Pressing the up and down keys will tilt the camera up and down. Experiment with panning and tilting to the furthest acceptable positions. The script will detect and reach these limits safely without going beyond the allowed servo angles. Press the spacebar at different angles to take snapshots at different camera positions. Figure 8.7 shows the message that is displayed when the tilt on the camera is moved upward.

Figure 8.7 Moving the camera: a text message is shown when a camera movement action is taken.

By creating this application, we have learned how to react to keyboard events to control our robot motor and camera hardware, perform motor movements, and capture images from the camera using different keyboard control keys.

Going Deeper: Robot kinematics

As you create robot projects that incorporate many servo motors acting as joints in a robotic arm, the field of robot kinematics has become increasingly important. By creating a model that incorporates the length of each link in the robotic arm and where each joint is, we can calculate at what angles to set each of our servos to move the robotic arm into different positions.

Forward kinematics work by taking the angle of each servo joint and then calculating where they would position the servo arm. Inverse kinematics work in reverse by taking a desired end position of the robotic arm and calculating what joint movements are required to get the arm into that position.

The Robotic Systems Guide (https://motion.cs.illinois.edu/RoboticSystems) by the University of Illinois at Urbana-Champaign provides an excellent reference on the topic of robot kinematics. Chapters 5 and 6 are dedicated to the topics of forward and inverse kinematics. The mathematical equations behind these calculations are presented with detailed visual diagrams of the robotic arms.

An interesting application of kinematics is creating a type of robot called a SCARA robot that only moves in the X–Y direction. This limitation makes the kinematics calculations simpler and requires fewer servo motors to create a functioning robot. The SCARA robot project (http://mng.bz/zOOB) on the Instructables website uses servo motors like the ones covered in this book to create this robot. Only two servo motors are required to move the robotic arm in the X–Y direction. One additional servo is used to lower and raise the pen for drawing operations.

Summary

- There is a dedicated servo to perform the tilt movement on the camera.
- The OpenCV library is used to interact with the Linux kernel video subsystem.
- The data structures in the NumPy library are heavily used in the OpenCV library.
- Placing text on an image is one of the common drawing operations to display a message in an OpenCV application.
- The Raspberry Pi camera will capture video images at the default frame rate of 30 frames per second.
- Unlike DC motors, we can move servo motors to exact positions because of their hardware sensors.

9 Face-following camera

This chapter covers

- Using the OpenCV library to detect faces in images
- Measuring and optimizing face detection performance
- Performing face detection in live videos
- Using servo motors to make a face-following camera

This chapter will first show how to use the OpenCV library to detect faces in images. Then, we will extend this functionality to detecting faces in a live video stream and measure and optimize our face detection process. Once we have a fast face detection mechanism in place, we will create an application to perform face detection in a live video stream. The last part of the chapter includes creating an application that can detect face movements and move the camera with motors in the direction of the detected face. Face detection is a demanding computer vision activity using machine learning to detect faces.

Machine learning plays an important role in the field of artificial intelligence, and it has many applications in robotics. In this chapter, we will create a robot that uses motors to move its camera based on the face it sees through image input data received from the camera. This is a powerful technique that can be extended to many robots that can automatically react and take action based on events detected in the environment. There are many autonomous robot systems created by taking sensor inputs robots are receiving, and they employ machine learning to decide what actions the robots should take to achieve their goals. These range from food delivery to construction, with robots preforming complex tasks such as bricklaying.

9.1 Hardware stack

Figure 9.1 shows the hardware stack, with the specific components used in this chapter highlighted. The robot will use a servo motor to move the attached camera in the direction of a detected face. Depending on whether the detected face is on the left or the right side of the camera, the servo will move the motor in the direction of the face. The initial applications in the chapter will focus on using the camera hardware to perform face detection, and later, the related servo movements will be added to the robot functionality.

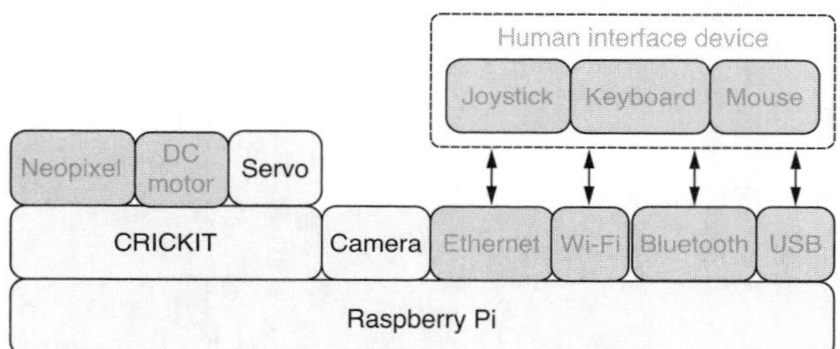

Figure 9.1 Hardware stack: servo motors will be used to move the camera toward a detected face.

9.2 Software stack

Details of the specific software used in this chapter are described in figure 9.2. We start the chapter by creating the `detect_face` application that will perform face detection using the OpenCV library on a single image. Then, we use the `measure_face` script and the `statistics` and `time` modules to measure the performance of our face detection process. Once we apply some performance enhancements, we will create the `face` library that can perform fast face detection, thus making the `live_face` application possible. The `live_face` application performs face detection in a live video stream. The chapter ends with the `follow` application that moves the servo motors to follow face movements. We will use the Linux video subsystem and camera hardware for the face detection. The `crickit` library will be used to control the servo motors.

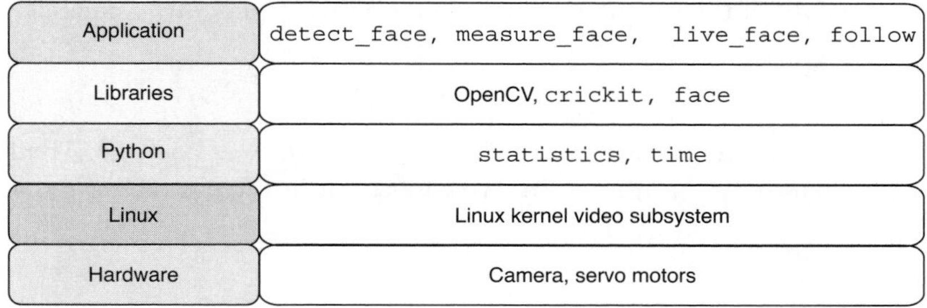

Figure 9.2 Software stack: the OpenCV library will be used to perform face detection.

9.3 Detecting faces in an image

The first step is to perform face detection on a single image. We need to create a Python application that meets the following requirements:

- The OpenCV computer vision library should be used to detect the location of a face in an image.
- The application should be able to draw a rectangle around the detected face and place a marker on its center.
- The `x,y` coordinates of the center of the face should be calculated and returned.

The final requirement of calculating the center of the face will be very helpful later in the chapter, as we will use it to decide in which direction to move the servo motor.

9.3.1 Exploring face detection

The OpenCV documentation (https://docs.opencv.org/4.x/) is an excellent source that provides good tutorials on common topics such as face detection. Under the Python tutorials section, it mentions that face detection is covered by the `objdetect` module. Specifically, the tutorial on the `objdetect` cascade classifier gives a detailed explanation of both theory and face detection application in OpenCV.

OpenCV performs face detection using the Haar feature-based cascade classifiers. This approach uses machine learning to train a cascade function from a large set of positive and negative images of faces. The positive images contain faces, and the negative images have no faces in them. Once the function is trained, we can then use it to detect faces in any image we provide.

Pretrained models are shipped as part of the OpenCV library and can be used directly. These models are XML files that can be found in the data directory of the OpenCV installation. We can start working with these models and performing face detection in the read–evaluate–print loop (REPL). The first step is to import the `cv2` package:

```
>>> import cv2
```

To locate the path of the OpenCV installation, we can inspect the `__path__` attribute:

```
>>> cv2.__path__
['/home/robo/pyenv/lib/python3.9/site-packages/cv2']
```

The `__path__` attribute provides a list of locations for the `cv2` package. The first item in the list is what we are interested in. We can save it in the `CV2_DIR` variable for further use:

```
>>> CV2_DIR = cv2.__path__[0]
>>> CV2_DIR
'/home/robo/pyenv/lib/python3.9/site-packages/cv2'
```

Now we can calculate the path of the model XML file for face detection and save it in a variable called `CLASSIFIER_PATH`:

```
>>> CLASSIFIER_PATH = f'{CV2_DIR}/data/haarcascade_frontalface_default.xml'
```

We can now create a classifier from the model file using the `CascadeClassifier` function. Once created, we save the classifier in a variable called `face_classifier`:

```
>>> face_classifier = cv2.CascadeClassifier(CLASSIFIER_PATH)
```

This classifier can be used to detect faces in images. Let's take the classifier for a spin and start detecting faces. Take a photo of a face with the camera, and save the image as `photo.jpg` in the same directory as the REPL session. We can open this image using the `imread` function:

```
>>> frame = cv2.imread('photo.jpg')
```

As expected, when we check the `shape` attribute of the image, we can see that the resolution of the image is 640 by 480 px with the three color components for each pixel. Figure 9.3 shows the image we are using in this REPL session:

```
>>> frame.shape
(480, 640, 3)
```

Figure 9.3 Image of face: an image of a face is taken with the Raspberry Pi camera.

Our classifier will check the pixel intensities in different regions of the image. To do this, you want the image to be represented as a grayscale image instead of color. We can convert our color image to grayscale by calling `cvtColor`:

```
>>> gray = cv2.cvtColor(frame, cv2.COLOR_BGR2GRAY)
```

If we inspect the `shape` attribute of our new `gray` image, we can see that it no longer has the three color components for each pixel. Instead, it has a single value for pixel intensity ranging from `0` to `255`, indicating a value from `0` for black, then higher values for gray, all the way to white at `255`.

```
>>> gray.shape
(480, 640)
```

The second operation we will perform on the image before face detection is histogram equalization. This operation improves the contrast in an image, which in turn improves the accuracy of our face detection. We save the prepared image into a variable called `clean`. Figure 9.4 shows what the resulting image will look like after we have applied histogram equalization:

```
>>> clean = cv2.equalizeHist(gray)
```

Figure 9.4 Equalized histogram: the contrast of the image is improved after histogram equalization.

We can now call the `detectMultiScale` method on our classifier that will perform face detection on our image and return the results as a list of detected faces:

```
>>> faces = face_classifier.detectMultiScale(clean)
```

When we inspect the length of `faces`, we can see that a single face was successfully detected in the image:

```
>>> len(faces)
1
```

Inspecting `faces` shows that for each detected face, a set of values is provided relating to that detected face. Each set of values relates to a matching rectangle:

```
>>> faces
array([[215, 105, 268, 268]])
```

We can save the rectangle values for the first detected face into variables indicating the top left coordinates `x, y` of the rectangle, as well as the variables `w, h` for the width and height of the rectangle:

```
>>> x, y, w, h = faces[0]
```

We can see the top-left corner of the matching face is located at coordinates `(215, 105)`:

```
>>> x, y
(215, 105)
```

We now have enough to slap together our first face detection application. Let's take what we have learned and bring it all together into a script to detect faces in an image.

Going deeper: Machine learning with OpenCV

The OpenCV documentation has a comprehensive Machine Learning Overview (https://docs.opencv.org/4.x/dc/dd6/ml_intro.html) that is a great launching point to delve deeper into the topic of machine learning in OpenCV.

At the heart of machine learning are algorithms that use training data to build and train models that can then make predictions based on that data. Once these trained models are in place, we can feed them new data that has not been seen before by the algorithms, and they can make predictions based on the data. In this chapter, we used a model that was trained on a set of face images to detect the presence and locations of faces in new images.

Another computer vision application is performing OCR (optical character recognition) on hand-written digits. The OpenCV project has samples of 5,000 handwritten digits that can be used as training data to train our models. The k-nearest neighbors algorithm can be used to train our models and then use them to recognize the digits in images. There is an excellent example of this in the Python tutorials of the OpenCV documentation under the Machine Learning section.

9.3.2 Marking detected faces

We will create a script to perform face detection on an image and then draw a rectangle around the matching face. We will also calculate the center of the matching rectangle and place a marker at the center point. Once we complete the detection and shape drawing, we will display the final image in our graphical application. The first step will be to import the `cv2` library:

```
import cv2
```

The value for the color blue is saved in the variable `BLUE`, and the location of the `cv2` library is saved in `CV2_DIR`. We can now set our `CLASSIFIER_PATH` by using `CV2_DIR`. Our face classifier is then created and saved in `face_classifier`:

```
BLUE = (255, 0, 0)
CV2_DIR = cv2.__path__[0]
CLASSIFIER_PATH = f'{CV2_DIR}/data/haarcascade_frontalface_default.xml'
face_classifier = cv2.CascadeClassifier(CLASSIFIER_PATH)
```

The `prep_face` function will prepare an image for face detection by converting it to grayscale and applying histogram equalization. The prepared image is then returned:

```
def prep_face(frame):
    gray = cv2.cvtColor(frame, cv2.COLOR_BGR2GRAY)
    return cv2.equalizeHist(gray)
```

We will then define `get_center` to calculate the center coordinates of a rectangle. We can use it to calculate the center of a detected face. The function receives the standard values relating to a rectangle and then returns the center point as an `x,y` coordinate pair:

```
def get_center(x, y, w, h):
    return int(x + (w / 2)), int(y + (h / 2))
```

The `detect_face` function receives an image and returns the center coordinates of a matching face. It first calls `prep_face` to prepare the image for face detection, and then `detectMultiScale` is called to detect faces in the image. If a face is found, we save the rectangle values of the first matching face into the variables `x, y, w, h`. Then, we calculate the center of the face and save this value in `center`. The `rectangle` function is used to draw a rectangle around the face, and `drawMarker` is used to place a marker at the center of the face. Finally, the coordinates of the center of the face are returned:

```
def detect_face(frame):
    clean = prep_face(frame)
    faces = face_classifier.detectMultiScale(clean)
    if len(faces) > 0:
        x, y, w, h = faces[0]
        center = get_center(x, y, w, h)
        cv2.rectangle(frame, (x, y), (x + w, y + h), BLUE, 2)
        cv2.drawMarker(frame, center, BLUE)
        return center
```

The `main` function loads our face image into a variable called `frame`. Then, `detect_face` is called to perform face detection, and the center of the face is saved in the `center` variable. These coordinates are printed out, and an image of the face is shown using `imshow`. The `waitKey` function is called to display the image until a key is pressed in the application:

```
def main():
    frame = cv2.imread('photo.jpg')
    center = detect_face(frame)
    print('face center:', center)
    cv2.imshow('preview', frame)
    cv2.waitKey()
```

The full script can be saved as `detect_face.py` on the Pi and then executed.

Listing 9.1 `detect_face.py`: Detecting a face and marking a matching face

```
#!/usr/bin/env python3
import cv2

BLUE = (255, 0, 0)
CV2_DIR = cv2.__path__[0]
CLASSIFIER_PATH = f'{CV2_DIR}/data/haarcascade_frontalface_default.xml'
face_classifier = cv2.CascadeClassifier(CLASSIFIER_PATH)

def get_center(x, y, w, h):
    return int(x + (w / 2)), int(y + (h / 2))

def prep_face(frame):
    gray = cv2.cvtColor(frame, cv2.COLOR_BGR2GRAY)
    return cv2.equalizeHist(gray)

def detect_face(frame):
    clean = prep_face(frame)
    faces = face_classifier.detectMultiScale(clean)
    if len(faces) > 0:
        x, y, w, h = faces[0]
        center = get_center(x, y, w, h)
        cv2.rectangle(frame, (x, y), (x + w, y + h), BLUE, 2)
        cv2.drawMarker(frame, center, BLUE)
        return center

def main():
    frame = cv2.imread('photo.jpg')
    center = detect_face(frame)
    print('face center:', center)
    cv2.imshow('preview', frame)
    cv2.waitKey()

main()
```

When this script is run, it will perform face detection on the `photo.jpg` image and draw the matching rectangle and marker around the detected face. Figure 9.5 shows what the application will look like once it has completed face detection and drawn shapes around the matching face.

Now that we have the groundwork in place for face detection in images, we can move on to the exciting task of performing face detection in a live video stream.

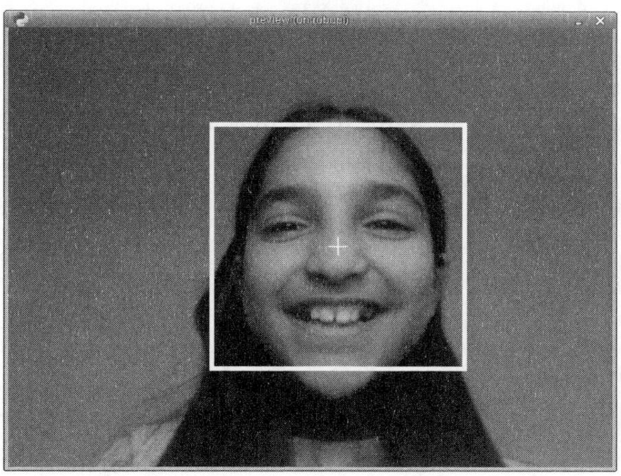

Figure 9.5 Face detection: rectangle and marker show the location of the detected face.

9.4 Detecting faces in live video

Detecting faces in live video follows a similar approach to detecting faces in a single image. The main difference is the more demanding performance requirements of doing face detection fast enough to keep up with a live video stream. We need to create a Python application that meets the following requirements:

- Face detection should be performed on each frame captured from the camera video stream.
- Face detection should be fast enough to keep up with the camera frame rate.
- The live video stream in the application should be displayed with any detected faces shown with a matching rectangle and marker.

The first task at hand will be to measure the performance of our face detection to see whether it is executed fast enough to keep up with the rate of the images we will receive from the video stream.

9.4.1 Measuring face detection performance

We know from the previous chapter that our camera will be capturing images at a rate of 30 frames per second. We need the face detection process to run faster than this frame rate so that it can keep up with the video stream. We will create a script to perform face detection multiple times and then report the average frame rate achieved for face detection.

The `cv2` library is imported to perform face detection. The `mean` function is imported to calculate the average frame rate. The `time` module will be used to measure the execution time of the face detection operation:

```
import cv2
from statistics import mean
import time
```

The functions and process of face detection are identical to what was used in the `detect_face.py` script. We will use the `get_detect_timing` function to measure the execution time of face detection. This function records the start time and then makes a call to the `detect_face` function. At the end, it calculates the elapsed time in seconds and returns the value:

```
def get_detect_timing(frame):
    start = time.perf_counter()
    center = detect_face(frame)
    return time.perf_counter() - start
```

Our `main` function will, as before, open the `photo.jpg` image and save it in `frame`. Then we make an initial call to `detect_face` and print out the coordinates of the center of the matching face. Next, we make repeated calls to `get_detect_timing` to capture a sample of 10 execution times. We take the average of this sample and calculate and report the average frames per second achieved. During each face detection, we use `frame.copy()` to provide a clean copy of the frame on each face detection:

```
def main():
    frame = cv2.imread('photo.jpg')
    center = detect_face(frame.copy())
    print('face center:', center)
    stats = [get_detect_timing(frame.copy()) for i in range(10)]
    print('avg fps:', 1 / mean(stats))
```

The full script can be saved as `measure_face.py` on the Pi and then executed.

Listing 9.2 `measure_face.py`: Measuring face detection performance

```
#!/usr/bin/env python3
import cv2
from statistics import mean
import time

BLUE = (255, 0, 0)
CV2_DIR = cv2.__path__[0]
CLASSIFIER_PATH = f'{CV2_DIR}/data/haarcascade_frontalface_default.xml'
face_classifier = cv2.CascadeClassifier(CLASSIFIER_PATH)

def get_center(x, y, w, h):
    return int(x + (w / 2)), int(y + (h / 2))

def prep_face(frame):
    gray = cv2.cvtColor(frame, cv2.COLOR_BGR2GRAY)
    return cv2.equalizeHist(gray)

def detect_face(frame):
    clean = prep_face(frame)
    faces = face_classifier.detectMultiScale(clean)
    if len(faces) > 0:
        x, y, w, h = faces[0]
        center = get_center(x, y, w, h)
```

```
        cv2.rectangle(frame, (x, y), (x + w, y + h), BLUE, 2)
        cv2.drawMarker(frame, center, BLUE)
        return center

def get_detect_timing(frame):
    start = time.perf_counter()
    center = detect_face(frame)
    return time.perf_counter() - start

def main():
    frame = cv2.imread('photo.jpg')
    center = detect_face(frame.copy())
    print('face center:', center)
    stats = [get_detect_timing(frame.copy()) for i in range(10)]
    print('avg fps:', 1 / mean(stats))

main()
```

When this script is run, it will perform face detection on the `photo.jpg` image. The coordinates of the center of the detected face are printed out in the terminal. Then, we take 10 samples of measuring the time required to detect a face. Based on the average of these samples, the frame rate is calculated and reported. We can see that the reported frame rate of 10.1 frames per second is well below the 30 frames per second that we need:

```
$ measure_face.py
face center: (349, 239)
avg fps: 10.104761447758944
```

Now that we have quantified our face detection performance, we can see that there is a performance problem, and we can get to work at improving the performance of our face detection process so that we can meet, and hopefully exceed, the 30 frames per second requirement.

9.4.2 Reducing the number of pixels to process

Our face detection operation does not need a large image to accurately detect faces. If we call our face classifier with a smaller image, it will have less pixels to process and will return the results faster. So the strategy we will take is to perform the face detection on an image that has been resized to be much smaller and thus processed faster.

We will resize the image to be 20% percent the size of the original image. We can find through experimentation that if this value is significantly smaller than 10%, it will affect detection accuracy. We will see that setting the value at 20% meets our performance needs and is within a safe range.

We can open a REPL session and do some calculations to get a sense of how much we have reduced the quantity of pixels in the image by doing this scaling. Our 20% scaling is equivalent to reducing the width and height of the image by a factor of 5. We can easily see this with the following calculation:

```
>>> 1/5
0.2
```

The captured image has a width of `640` and a height of `480`. We can calculate the height and width of the scaled-down image with the following calculations:

```
>>> 640/5
128.0
>>> 480/5
96.0
```

We can see that the resized image will have a width of `128` and a height of `96`. We can now calculate the total number of pixels in the original and the resized image:

```
>>> 640*480
307200
>>> 128*96
12288
```

Now, we can take these two pixel counts and divide them to find out by what factor we have reduced the total number of pixels:

```
>>> 307200/12288
25.0
```

We have reduced the total number of pixels to process by a 25x factor. This is a big reduction in data to process and should yield a big improvement in processing speed. Figure 9.6 shows the significant difference in image sizes when we place the two images side by side. We can cross-check this figure by squaring the reduction factor on the width and height:

```
>>> 5*5
25
25.0
```

As expected, it produced the same result of a 25x reduction factor.

Figure 9.6 Image reduction: the original and reduced images are kept side by side for comparison.

We can take the smaller image and run it through our face detection script to inspect the results. Figure 9.7 shows that the image is clearly pixelated, but this doesn't pose a problem for the face detection procedure.

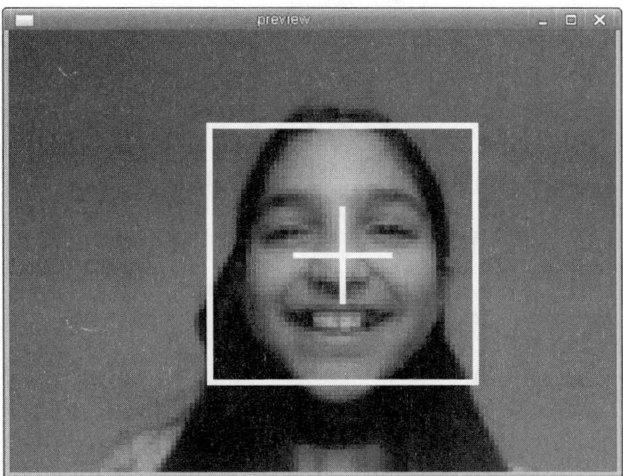

Figure 9.7 Detection on smaller images: face detection is successful for images with smaller resolution.

Now that we have our initial calculations out of the way, we can move on to implement the new faster version of our application.

9.4.3 Optimizing face detection performance

This implementation will build on the previous one and mainly add the image reduction steps to gain a performance boost. First, we will import the `cv2` library to perform face detection:

```
import cv2
```

The scaling factor is saved in the `DETECT_SCALE` variable:

```
DETECT_SCALE = 0.2
```

The `resize` function receives the image and the desired scale to resize the image and return the new smaller image. The new width and height of the image are calculated based on the provided `scale` and saved in `size`. The `cv2.resize` function is then called on the image. The OpenCV documentation (https://docs.opencv.org/4.x) on the resize function gives guidance to use the `INTER_AREA` interpolation when shrinking an image and `INTER_CUBIC` when enlarging it. We are shrinking the image, so we use `INTER_AREA`:

```
def resize(img, scale):
    size = (int(img.shape[1] * scale), int(img.shape[0] * scale))
    return cv2.resize(img, size, interpolation=cv2.INTER_AREA)
```

The `detect_face` function now has a performance enhancement. After `prep_face` is called, a call is made to `resize` to create a smaller image before face detection. Then, `detectMultiScale` is called using `small`. When the rectangle values are returned, we divide them by `DETECT_SCALE` so that they can be mapped again to the original full-resolution image. In this way, we can show details of the detected face on the full-sized original image but obtain a performance gain by doing face detection on a smaller image. The rest of the code remains the same:

```
def detect_face(frame):
    clean = prep_face(frame)
    small = resize(clean, DETECT_SCALE)
    faces = face_classifier.detectMultiScale(small)
    if len(faces) > 0:
        x, y, w, h = [int(i / DETECT_SCALE) for i in faces[0]]
        center = get_center(x, y, w, h)
        cv2.rectangle(frame, (x, y), (x + w, y + h), BLUE, 2)
        cv2.drawMarker(frame, center, BLUE)
        return center
```

The library can be saved as `face.py` on the Pi to be imported by other applications.

Listing 9.3 `face.py`: Providing a fast face detection library

```
import cv2

BLUE = (255, 0, 0)
CV2_DIR = cv2.__path__[0]
CLASSIFIER_PATH = f'{CV2_DIR}/data/haarcascade_frontalface_default.xml'
face_classifier = cv2.CascadeClassifier(CLASSIFIER_PATH)
DETECT_SCALE = 0.2

def resize(img, scale):
    size = (int(img.shape[1] * scale), int(img.shape[0] * scale))
    return cv2.resize(img, size, interpolation=cv2.INTER_AREA)

def get_center(x, y, w, h):
    return int(x + (w / 2)), int(y + (h / 2))

def prep_face(frame):
    gray = cv2.cvtColor(frame, cv2.COLOR_BGR2GRAY)
    return cv2.equalizeHist(gray)

def detect_face(frame):
    clean = prep_face(frame)
    small = resize(clean, DETECT_SCALE)
    faces = face_classifier.detectMultiScale(small)
    if len(faces) > 0:
        x, y, w, h = [int(i / DETECT_SCALE) for i in faces[0]]
        center = get_center(x, y, w, h)
        cv2.rectangle(frame, (x, y), (x + w, y + h), BLUE, 2)
        cv2.drawMarker(frame, center, BLUE)
        return center
```

To see this library in action, we will create a new script that will import the library and make multiple calls to the face detection functions and measure their performance. We first import `cv2`, `mean`, and `time` as we have done before to open images, calculate averages, and measure execution times. We then import the `detect_face` function from our new `face` library:

```
import cv2
from face import detect_face
from statistics import mean
import time
```

The rest of the application has the same functions as those created in the `measure_face.py` script to measure execution time and report the average frame rate achieved.

The full script can be saved as `fast_face.py` on the Pi and then executed.

Listing 9.4 `fast_face.py`: Reporting performance of the fast face detection functions

```
#!/usr/bin/env python3
import cv2
from face import detect_face
from statistics import mean
import time

def get_detect_timing(frame):
    start = time.perf_counter()
    center = detect_face(frame)
    return time.perf_counter() - start

def main():
    frame = cv2.imread('photo.jpg')
    center = detect_face(frame.copy())
    print('face center:', center)
    stats = [get_detect_timing(frame.copy()) for i in range(10)]
    print('avg fps:', 1 / mean(stats))

main()
```

When this script is run, it will call our new faster face detection implementation. We can see from the results that a big performance gain has been achieved, as now we have reached a frame rate of 75.6 frames per second. This gives us a performance boost that is over seven times faster than the previous approach to face detection:

```
$ fast_face.py
face center: (347, 242)
avg fps: 75.63245789951259
```

This frame rate is also way above the 30 frames per second target that we were hoping to achieve. We can now proceed to use this new improved approach to face detection in a live video stream.

9.4.4 Showing detected faces in live video

In the following script, we will capture images from the camera video stream face detection and then show the detected faces in an application window. The `cv2` library is imported to capture images from the camera video stream. The `detect_face` function is imported to perform face detection:

```
import cv2
from face import detect_face
```

As we have done before, the key code for the Esc key is saved in `ESC_KEY`. It will be used to exit the graphical application by pressing the Esc key:

```
ESC_KEY = 27
```

The `main` function will save a video capture object in the variable `cap`. We then check whether the capture device was opened correctly. We enter an event loop that we continue looping in until the Esc or Q key is pressed. In each loop iteration, we capture a frame from the video stream and call the `detect_face` function on the captured image. We then call `imshow` to show the captured image with any detected faces marked. When this loop is exited, we release the video capture device by calling the `cap.release` function:

```
def main():
    cap = cv2.VideoCapture(0)
    assert cap.isOpened(), 'Cannot open camera'
    while cv2.waitKey(1) not in [ord('q'), ESC_KEY]:
        ret, frame = cap.read()
        assert ret, 'Cannot read frame from camera'
        detect_face(frame)
        cv2.imshow('preview', frame)
    cap.release()
```

The full script can be saved as `live_face.py` on the Pi and then executed.

Listing 9.5 `live_face.py`: Showing detected faces in a live video stream

```
#!/usr/bin/env python3
import cv2
from face import detect_face

ESC_KEY = 27

def main():
    cap = cv2.VideoCapture(0)
    assert cap.isOpened(), 'Cannot open camera'
    while cv2.waitKey(1) not in [ord('q'), ESC_KEY]:
        ret, frame = cap.read()
        assert ret, 'Cannot read frame from camera'
        detect_face(frame)
        cv2.imshow('preview', frame)
    cap.release()

main()
```

When this script is run, it will continually capture images from the video stream. Each image is passed through our face detection functions. If a face is detected, a rectangle is drawn around the detected face with a marker placed at its center. The video stream with face detection is shown in the application window until the Esc key or Q key is pressed to exit the application.

Figure 9.8 shows what the camera looks like once it is mounted on the Pan-Tilt Kit. The two servo motors give the camera the ability to be moved in different directions.

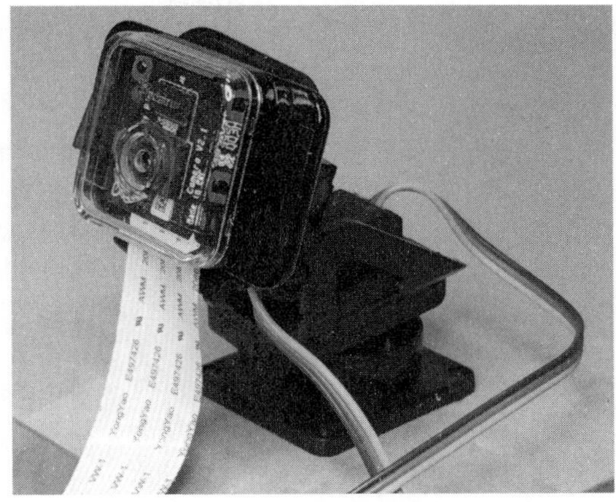

Figure 9.8 Camera on Pan-Tilt Kit: the camera is mounted on the kit to enable camera movements.

In the next section, we will use the servo motor to move the camera in the direction of the detected face.

9.5 Creating a face-following robot

Now that we can detect faces fast enough to process live video, we can take our code to the next level and have the robot react to where you position your face. The robot will move the camera to follow your face. We need to create a Python application that meets the following requirements:

- It should be able to recognize whether a face is detected in the left, center, or right of frame.
- When the face is detected in the left or right, the camera should move toward the face.
- In the application, a live video stream with detected faces marked and a grid showing the three zones (left, center, and right) should be displayed.

Showing the three zones in the application will make the application more interactive, as people will be able to tell which zone their face has been detected in and where the camera will move.

9.5.1 Zoning the face detection

We can split up what the robot sees into three areas or zones. When a face is detected in the center zone, we don't have to do anything because the camera is facing the person. If the face is detected in the left zone, then we will move the servo so that the camera places the face in the center zone. If the face is detected in the right zone, we will move the servo again but in the opposite direction. We will focus only on moving the camera left and right using the pan movements of the servo motor. Figure 9.9 shows the three zones for face detection.

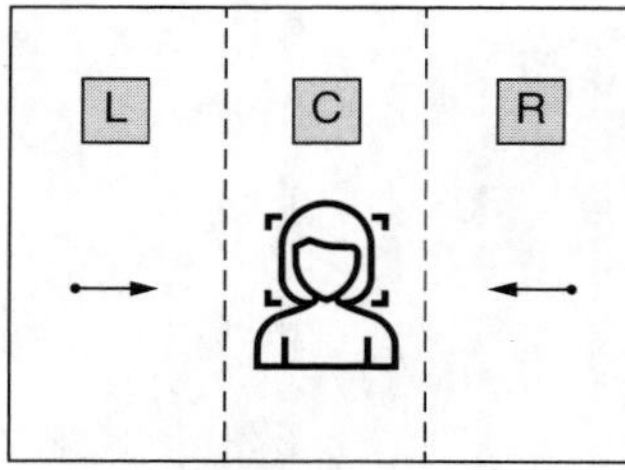

Figure 9.9 Camera zones: the camera area is split into three detection zones.

Now, let's pop into a REPL session and see how we can split the camera viewing area into these three zones. First, we will import `cv2` for drawing on images and `numpy` for creating a new blank image:

```
>>> import cv2
>>> import numpy as np
```

We will save the width and height of our camera images into the variables `IMG_WIDTH` and `IMG_HEIGHT`. This will make our code more readable:

```
>>> IMG_WIDTH = 640
>>> IMG_HEIGHT = 480
```

We can get the center or halfway point of the width by dividing `IMG_WIDTH` by two:

```
>>> (IMG_WIDTH / 2)
320.0
```

Now, let's take this center position and move left 50 px to get the position of the line between the left and center zones. We will save this value in a variable called `LEFT_X`:

```
>>> LEFT_X = int((IMG_WIDTH / 2) - 50)
>>> LEFT_X
270
```

By moving right 50 px from the center, we get the position of the line between the center and right zones. We save this value in `RIGHT_X`:

```
>>> RIGHT_X = int((IMG_WIDTH / 2) + 50)
>>> RIGHT_X
370
```

We can save the value for the color green in a variable called `GREEN`:

```
>>> GREEN = (0, 255, 0)
```

Next, let's create a blank color image with our desired dimensions:

```
>>> img = np.zeros(shape=(480, 640, 3), dtype=np.uint8)
```

We can draw a grid showing the three zones by drawing a rectangle around the center zone:

```
>>> cv2.rectangle(img, (LEFT_X, -1), (RIGHT_X, IMG_HEIGHT), GREEN)
```

The last step will be to save what we have created so that we can see the image. We will use `imwrite` to save the image as the filename `zones.jpg`:

```
>>> cv2.imwrite('zones.jpg', img)
```

Figure 9.10 shows what the image will look like once we have drawn the zone grid. The center zone is set to be narrower than the left and right zones. In this way, we can make the camera more sensitive to moving left and right when the face moves around the frame.

Figure 9.10 Zone grid: the zone grid is drawn using the rectangle method.

9.5.2 Moving motors to follow faces

We can now have a go at tackling the script to follow the person's face as they look at different zones within the camera's viewing area. We can build on the experimentation we did in the previous section.

We import the `cv2` library to capture images from the camera. The `detect_face` function is imported and will perform face detection as we have seen before. Finally, we use the `crickit` module to control the servo motor that has a camera attached to it:

```
import cv2
from face import detect_face
from adafruit_crickit import crickit
```

Next, we define `ESC_KEY` and `GREEN` to store the key code for the Esc key and the value for the color green. The image height and width are defined in `IMG_WIDTH` and `IMG_HEIGHT`. We then calculate the values for `LEFT_X` and `RIGHT_X` to help keep track of the zone the face is detected in:

```
ESC_KEY = 27
GREEN = (0, 255, 0)
IMG_WIDTH = 640
IMG_HEIGHT = 480
LEFT_X = int((IMG_WIDTH / 2) - 50)
RIGHT_X = int((IMG_WIDTH / 2) + 50)
```

As we have done in chapter 8, we create a variable called `PAN` to keep track of values relating to our servo that performs the pan movement. Namely, we keep a reference to the servo object to the minimum, maximum, and starting angles of the servo. We also keep the actuation range setting in `range`. As done in the previous chapter, we store the value of the angle changes at each step in `ANGLE_STEP`. We use `MOVE` to map the left, center, and right zones to their associated servo movements:

```
PAN = dict(servo=crickit.servo_1, min=30, max=110, start=70, range=142)
ANGLE_STEP = 2
MOVE = dict(L=ANGLE_STEP, C=0, R=-ANGLE_STEP)
```

The `get_zone` function will return the zone of the detected face based on the values of `LEFT_X` and `RIGHT_X`:

```
def get_zone(face_x):
    if face_x <= LEFT_X:
        return 'L'
    elif face_x <= RIGHT_X:
        return 'C'
    else:
        return 'R'
```

The `init_motors` function is used to initialize the servo motor's starting position and actuation range:

```
def init_motors():
    PAN['servo'].actuation_range = PAN['range']
    PAN['servo'].angle = PAN['start']
```

We will use the `move_motor` function to move the servo based on the position of the detected face. We first calculate the zone by calling `get_zone`. Then, we look up the angle change and save it in `change`. Next, we apply the new angle if a change is detected and if the new angle falls within our minimum and maximum angle range:

```
def move_motor(face_x):
    zone = get_zone(face_x)
    change = MOVE[zone]
    if change and PAN['min'] <= PAN['servo'].angle + change <= PAN['max']:
        PAN['servo'].angle += change
```

When we create a new video capture object, we call `check_capture_device` to check the device. We check whether it was opened successfully and whether the width and height of images being captured by the device match our `IMG_WIDTH` and `IMG_HEIGHT` values:

```
def check_capture_device(cap):
    assert cap.isOpened(), 'Cannot open camera'
    assert cap.get(cv2.CAP_PROP_FRAME_WIDTH) == IMG_WIDTH, 'wrong width'
    assert cap.get(cv2.CAP_PROP_FRAME_HEIGHT) == IMG_HEIGHT, 'wrong height'
```

The `main` function first calls `init_motors` to initialize the servo motors. Then we create a video capture device and check it by calling `check_capture_device`. We then enter an event loop that we only exit if either the Esc or the Q key is pressed. In each loop, we grab an image from the video stream and save it in `frame`. We then call `detect_face` to perform face detection and return us to the position of the center of the detected face. If a face was detected, we call `move_motor` with the `x` coordinate of the detected face. We then draw our zone grid onto the image by calling `cv2.rectangle` with the related dimensions. The last step of the loop is to show the latest video frame in the application by calling `imshow`. When we exit the loop, we call `cap.release` to release the video capture device:

```
def main():
    init_motors()
    cap = cv2.VideoCapture(0)
    check_capture_device(cap)
    while cv2.waitKey(1) not in [ord('q'), ESC_KEY]:
        ret, frame = cap.read()
        assert ret, 'Cannot read frame from camera'
        center = detect_face(frame)
        if center:
            move_motor(center[0])
        cv2.rectangle(frame, (LEFT_X, -1), (RIGHT_X, IMG_HEIGHT), GREEN)
        cv2.imshow('preview', frame)
    cap.release()
```

The full script can be saved as `follow.py` on the Pi and then executed.

Listing 9.6 `follow.py`: Moving the camera to follow detected faces

```
#!/usr/bin/env python3
import cv2
from face import detect_face
from adafruit_crickit import crickit

ESC_KEY = 27
GREEN = (0, 255, 0)
IMG_WIDTH = 640
IMG_HEIGHT = 480
LEFT_X = int((IMG_WIDTH / 2) - 50)
RIGHT_X = int((IMG_WIDTH / 2) + 50)
PAN = dict(servo=crickit.servo_1, min=30, max=110, start=70, range=142)
ANGLE_STEP = 2
MOVE = dict(L=ANGLE_STEP, C=0, R=-ANGLE_STEP)

def get_zone(face_x):
    if face_x <= LEFT_X:
        return 'L'
    elif face_x <= RIGHT_X:
        return 'C'
    else:
        return 'R'

def move_motor(face_x):
    zone = get_zone(face_x)
    change = MOVE[zone]
    if change and PAN['min'] <= PAN['servo'].angle + change <= PAN['max']:
        PAN['servo'].angle += change

def init_motors():
    PAN['servo'].actuation_range = PAN['range']
    PAN['servo'].angle = PAN['start']

def check_capture_device(cap):
    assert cap.isOpened(), 'Cannot open camera'
```

```
    assert cap.get(cv2.CAP_PROP_FRAME_WIDTH) == IMG_WIDTH, 'wrong width'
    assert cap.get(cv2.CAP_PROP_FRAME_HEIGHT) == IMG_HEIGHT, 'wrong height'

def main():
    init_motors()
    cap = cv2.VideoCapture(0)
    check_capture_device(cap)
    while cv2.waitKey(1) not in [ord('q'), ESC_KEY]:
        ret, frame = cap.read()
        assert ret, 'Cannot read frame from camera'
        center = detect_face(frame)
        if center:
            move_motor(center[0])
        cv2.rectangle(frame, (LEFT_X, -1), (RIGHT_X, IMG_HEIGHT), GREEN)
        cv2.imshow('preview', frame)
    cap.release()

main()
```

When this script is run, you can look into the camera and see your face in the live camera feed with a border placed around your detected face. The center of the face is also marked on the live image with a crosshair. From this marker, we can tell which zone the face is in. If you move the face outside the center zone, the servo will automatically reposition the camera to put your face back in this zone. Figure 9.11 shows a face that has been detected and marked in the left zone, which then makes the servo motors move the cameras to place the face back in the center zone.

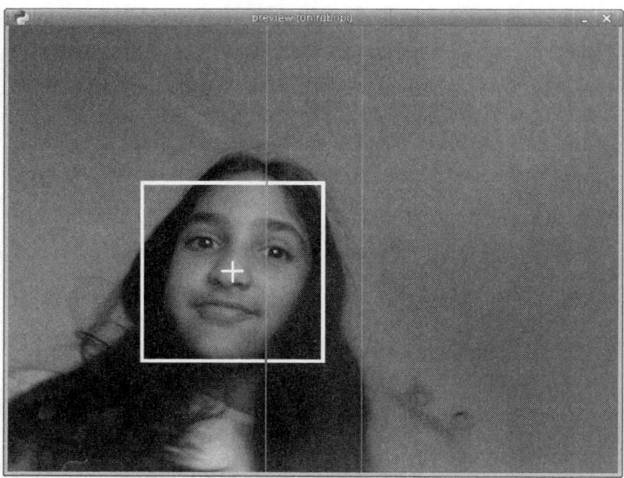

Figure 9.11 Zoned face: the face is detected and marked in the left zone.

This application gave us a chance to apply machine learning to our robotics project by using computer vision and face tracking. In the coming chapters, we will use other computer vision functionalities, such as QR code detection, to help our robot interact even further with its environment by using the camera as a way to see their environment.

Robots in the real world: Robotics vision processing

Robots can use computer vision to do feature detection to extract visual features such as corners and edges of objects. With this feature detection in place, robots can detect and classify objects they see in their environment.

One application of this object interaction is creating robots that can pick and place objects in manufacturing and logistics. Using computer vision, they identify an object, grab it, and then move it from one location to another.

Inspection robots are another case that combines computer vision and robotics to create robots that can be used as part of the quality control process in manufacturing to perform fully automated inspections of manufactured products.

Summary

- A fast face detection mechanism is required to perform face detection in a live video stream.
- Servo motors are used to move the attached camera in the direction of a detected face.
- Haar feature-based cascade classifiers are used in OpenCV to perform face detection.
- Histogram equalization improves the contrast of an image and helps to improve the accuracy of face detection.
- Detected faces will have a matching rectangle and marker drawn around them.
- The face detection must be able to handle a camera image rate of at least 30 frames per second for live face recognition to work.
- Calling the face classifier with smaller images can make face detection faster.

10 Robotic QR code finder

This chapter covers

- Generating QR codes
- Detecting and decoding data in QR codes
- Streaming live video using Motion JPEG
- Creating a robot that can search for specific QR codes in its environment

We start this chapter by exploring the QR code standard and learning how to generate our own QR codes. Then, we use the OpenCV computer vision library to detect QR codes in images, as well as to read the data encoded in the QR code itself. We will then learn how to save the video stream data from the camera to the filesystem so that multiple applications can simultaneously read live video data. This will allow us to check the video stream for QR codes we are interested in, as well as stream the video to desktop and web applications at the same time. We will use the Tornado web framework to create a Motion JPEG video server that can be accessed from any mobile device or computer on the network to get a live view of the robot's camera video stream. Finally, we end the chapter by creating a robot that can move around in search of matching QR codes in its environment.

Bringing all these technologies together helps us solve the problem of having robots use computer vision to investigate their environment and move around to different desired locations by looking for matching QR codes. This is a core functionality for many robots that must perform autonomous navigation in warehouses or factories.

10.1 Hardware stack

Figure 10.1 shows the hardware stack, with the specific components used in this chapter highlighted. The robot will use the DC motors to move back and forth along a set track. The camera is mounted on the side of the robot and can capture objects next to the robot as the robot drives past them. The robot will be checking the images from the live video feed, looking for a matching QR code. Once the code is found, the robot can stop the motors, as it has reached its desired destination.

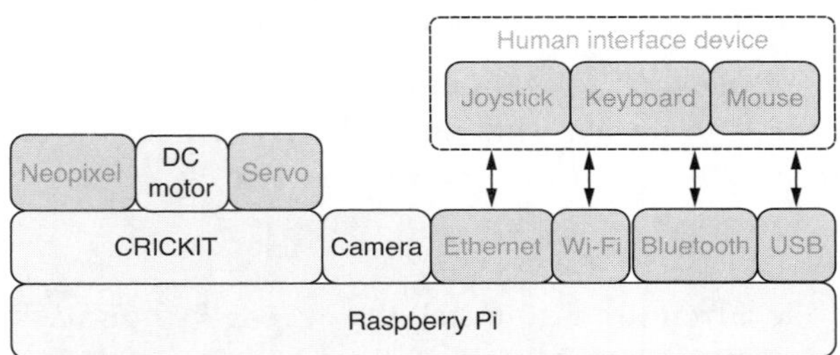

Figure 10.1 Hardware stack: the camera will be used to capture live images for QR code detection.

For further details on the robot assembly, check the robot assembly guide in appendix C. It shows how to assemble the robot used in this chapter. It also gives tips on how to create a track for the robot so that it can travel back and forth on a controlled path.

10.2 Software stack

Details of the specific software used in this chapter are described in figure 10.2. We start the chapter by creating the `detect_qr` application that will perform QR code detection and decoding using the OpenCV library on a single image. Then, we will

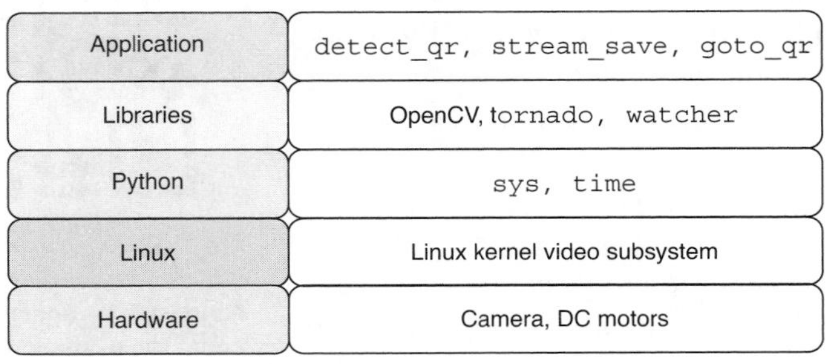

Figure 10.2 Software stack: the OpenCV library will be used to perform QR code detection.

use the `stream_save` script to capture the video stream to the filesystem. The `watcher` library uses the `sys` module to read command line arguments and watch for changes on the streaming image file. We then create streaming applications for both web and graphical applications. We end the chapter by using the camera and DC motors hardware in the `goto_qr` application to move the robot to a specific target location marked by a QR code.

10.3 Detecting QR codes in an image

The first step is to perform QR code detection and decoding on a single image. We need to create a Python application that meets the following requirements:

- The application should use the OpenCV computer vision library to detect the location of a QR code in an image.
- A rectangle should be drawn in the image around the detected QR code.
- The application should decode and return the data stored in the QR code.

The last requirement of decoding data stored in the QR code will be very helpful later in the chapter, as we will use it to decide whether we have reached our desired QR code or whether the robot should keep moving down the track.

10.3.1 Exploring QR codes

The first step in our QR code adventure is to install the `qrcode` Python package. This module will let us generate QR codes. Run the following command to install the package:

```
$ ~/pyenv/bin/pip install qrcode
```

This package can be imported in Python code or executed directly in the command line. Let's start by generating some QR codes in the terminal. When we run the next command, a QR code encoded with the text `hello` will be generated and outputted to our terminal:

```
$ ~/pyenv/bin/qr hello
```

You can test the QR code by scanning it using a smartphone. Once you scan the QR code, the text `hello` should appear on your device. When we run the next command, it will save the generated QR code to an image instead of outputting it to the terminal:

```
$ ~/pyenv/bin/qr hello > hello.png
```

You can open the `hello.png` image and test it again. This is a useful way for generating QR codes, as we can print the image and stick it to the objects we want to tag with QR codes. Figure 10.3 shows the QR code generated using this command.

Figure 10.3 Generated QR code: the example QR code has the text `hello` encoded in it.

Next, we will pop into a read–evaluate–print loop (REPL) session to explore using the package from a Python application. The first step is to import the `qrcode` library:

```
>>> import qrcode
```

The next line will create a QR code with the text `hi again` and save it into an image called `hi_again.png`:

```
>>> qrcode.make('hi again').save('hi_again.png')
```

For more advanced applications, we use a `QRCode` object. Through this object, we can set different options relating to the error correction, box size, and border of the QR code. In the example that follows, we create a `QRCode` object and use the `add_data` method to set the content of the QR code:

```
>>> qr = qrcode.QRCode()
>>> qr.add_data('www.python.org')
```

Next, we call the `make` method to generate the QR code. Once generated, we can obtain details about the generated QR code, such as the symbol version that was used. In this example, the `version` attribute reports the QR code is using symbol version `1`:

```
>>> qr.make()
>>> qr.version
1
```

The official website for the QR code standard (https://www.qrcode.com) gives exact details on each symbol version and how much data it can store. Essentially, the more data you put in a QR code, the larger the symbol version, which in turn generates a denser QR code. It is useful to be able to check this value when we generate QR code images, as the lower-version numbers are less dense and will be easier to read even with low-resolution images. The next line in our REPL will save the QR code to an image called `python.png`:

```
>>> qr.make_image().save('python.png')
```

On most smartphones, if you read this QR code, it will be detected from the text as a URL, and you will be directed to the Python website (https://www.python.org). Now that we have generated some QR codes, let's move on to detecting and decoding them.

10.3.2 Marking detected QR codes

We will create a script to perform QR code detection and decoding on an image and then to draw a rectangle around the matching QR code. We will also display the decoded data as text on the image. The image will then be displayed in our graphical application.

Take the image generated in the previous section that has the text `hello` encoded in it, and print it out. Then, take a photo of it using the Pi camera, and save the image in a file called `hello.jpg`. You can use the `snapshot.py` application from chapter 8 to take the image. Alternatively, a `hello.jpg` image is also provided in the folder for this chapter in the book's GitHub repository.

The first step will be to import the `cv2` library:

```
import cv2
```

The value for the color blue is saved in the variable `BLUE`, and the font for displaying text in the application is saved in `FONT`. We then instantiate a `QRCodeDetector` object and save it in `decoder`. Methods on this object will be called to perform the QR code detection and decoding:

```
BLUE = (255, 0, 0)
FONT = cv2.FONT_HERSHEY_SIMPLEX
decoder = cv2.QRCodeDetector()
```

The `draw_box` function will draw a box around the detected QR code in the image. The image is provided in the `frame` argument, and the four points of the detected quadrangle are provided in `points`. The color and thickness of the box are set using the `color` and `thickness` arguments. The point values are first converted to integers, as this is what is expected by the `cv2.line` function. We then save each of the four points of the quadrangle into its own variable. Next, we draw four lines to connect these four points:

```
def draw_box(frame, points, color, thickness):
    points = [(int(x), int(y)) for x, y in points]
    pt1, pt2, pt3, pt4 = points
    cv2.line(frame, pt1, pt2, color, thickness)
    cv2.line(frame, pt2, pt3, color, thickness)
    cv2.line(frame, pt3, pt4, color, thickness)
    cv2.line(frame, pt4, pt1, color, thickness)
```

We will then define `decode_qrcode`, which calls the `detectAndDecode` method to detect and decode QR codes in the `frame` image. Decoded data is stored in the variable `data`, and a list of matching points is stored in `matches`. If we find decoded data, we display it as text using `putText` and draw a box around the matched area by calling the `draw_box` function. We finally end the function by returning the decoded data:

```
def decode_qrcode(frame):
    data, matches, _ = decoder.detectAndDecode(frame)
    if data:
        cv2.putText(frame, f'data: {data}', (30, 30), FONT, 1, BLUE)
        draw_box(frame, matches[0], BLUE, thickness=3)
    return data
```

The `main` function loads our photo of a QR code into a variable called `frame`. Then `decode_qrcode` is called to perform the QR code detection and decoding. The decoded data is stored in a variable called `decoded_data` and printed out. The image is then shown using `imshow`. The `waitKey` function is called to display the image until a key is pressed in the application:

```
def main():
    frame = cv2.imread('hello.jpg')
    decoded_data = decode_qrcode(frame)
```

```
    print('decoded_data:', repr(decoded_data))
    cv2.imshow('preview', frame)
    cv2.waitKey()
```

The full script can be saved as `detect_qr.py` on the Pi and then executed.

Listing 10.1 `detect_qr.py`: Detecting and decoding QR codes in an image

```
#!/usr/bin/env python3
import cv2

BLUE = (255, 0, 0)
FONT = cv2.FONT_HERSHEY_SIMPLEX
decoder = cv2.QRCodeDetector()

def draw_box(frame, points, color, thickness):
    points = [(int(x), int(y)) for x, y in points]
    pt1, pt2, pt3, pt4 = points
    cv2.line(frame, pt1, pt2, color, thickness)
    cv2.line(frame, pt2, pt3, color, thickness)
    cv2.line(frame, pt3, pt4, color, thickness)
    cv2.line(frame, pt4, pt1, color, thickness)

def decode_qrcode(frame):
    data, matches, _ = decoder.detectAndDecode(frame)
    if data:
        cv2.putText(frame, f'data: {data}', (30, 30), FONT, 1, BLUE)
        draw_box(frame, matches[0], BLUE, thickness=3)
    return data

def main():
    frame = cv2.imread('hello.jpg')
    decoded_data = decode_qrcode(frame)
    print('decoded_data:', repr(decoded_data))
    cv2.imshow('preview', frame)
    cv2.waitKey()

main()
```

When this script is run, it will perform the QR code detection on the `hello.jpg` image and draw a box around the matching quadrangle. The decoded data is also displayed in the top-left corner of the image. Figure 10.4 shows what the application will look like once it has completed detection and decoding of the QR code.

Figure 10.4 Detected QR code: the application draws a box around the detected QR code.

We now have a solid foundation in QR code detection and decoding. We will have multiple applications, all wanting to access the live video stream. Thus, the next step will be to design a system to capture and distribute images from the live video stream to multiple applications.

10.4 Streaming live video to multiple applications

We will approach this problem by saving the images from the live video stream to the filesystem. Then, multiple applications can simultaneously read these images from the filesystem and use them to stream to desktop or web applications. We can use the same mechanism to detect QR codes in the live stream and control the robot's movements. We need to create a Python application that meets the following requirements:

- The latest frame from the camera video stream should be captured and saved to the filesystem.
- The frame should be saved to a ramdisk in order to not create any additional disk workload.
- The image data should be saved as an atomic operation to ensure data consistency.

The first step is to create an application to save the frames from the video stream to the filesystem. We can then create applications to stream the video stream to desktop and web applications. By using a ramdisk, we will get better I/O performance for video streaming and won't create a slowdown for other applications that are trying to read and write from the disk.

10.4.1 Saving the video stream to a ramdisk

One thing to keep in mind is that the camera on the robot is positioned upside down so that there can be enough space for the camera connector to be connected to the Raspberry Pi. This will make our captured images appear upside down. This problem can be solved by correcting the image orientation in software and flipping the image after we capture it.

The `cv2` library is imported to capture frames from the video stream. The `os` module is imported so that we can access environment variables:

```
import os
import cv2
```

A ramdisk is created by default on Linux systems, and we can access its location by reading the value of the `XDG_RUNTIME_DIR` environment variable. The files in the ramdisk are stored in memory and not on physical disks. In this way, we can work with them as we would with any other file on the filesystem but get the added benefit of not putting any additional load on the physical disks. We will place our image in this directory and use the `IMG_PATH` variable to keep track of its path. We also need to save the image data to a temporary file located in `TMP_PATH` before saving it to its final location:

```
IMG_PATH = os.environ['XDG_RUNTIME_DIR'] + '/robo_stream.jpg'
TMP_PATH = os.environ['XDG_RUNTIME_DIR'] + '/robo_stream_tmp.jpg'
```

We will set the sizes of the images we capture to be half the size of the default resolution. This will make the size of data saved and streamed smaller and more efficient. The image will still be large enough to get a good view of what the robot sees as well as accurately perform QR code detection and decoding. We save these values in the variables `FRAME_WIDTH` and `FRAME_HEIGHT`:

```
FRAME_WIDTH = 320
FRAME_HEIGHT = 240
```

The `init_camera` function is used to create the video capture object and set the video capture dimensions to `FRAME_WIDTH` and `FRAME_HEIGHT`. The video capture object is then returned:

```
def init_camera():
    cap = cv2.VideoCapture(0)
    assert cap.isOpened(), 'Cannot open camera'
    cap.set(cv2.CAP_PROP_FRAME_WIDTH, FRAME_WIDTH)
    cap.set(cv2.CAP_PROP_FRAME_HEIGHT, FRAME_HEIGHT)
    return cap
```

The `save_frames` function enters into an infinite loop and captures a frame from the video stream in each loop. The variable `counter` keeps track of the number of frames captured so far. We save the captured image in `frame` and then flip the image by calling `cv2.flip`. We use `imwrite` to save the image to our temporary file. Then, we call `os.replace` to place our temporary file in its final destination. This call is guaranteed to be an atomic operation on Unix operating systems such as Linux, which our Pi is running on. Then, we print out how many frames we have captured so far. We use the carriage return as the end character when printing the output so that the same line in the terminal gets updated with our frame counter:

```
def save_frames(cap):
    counter = 0
    while True:
        counter += 1
        ret, frame = cap.read()
        assert ret, 'Cannot read frame from camera'
        frame = cv2.flip(frame, -1)
        cv2.imwrite(TMP_PATH, frame)
        os.replace(TMP_PATH, IMG_PATH)
        print('frames:', counter, end='\r', flush=True)
```

Finally, we end with our `main` function that first initializes the video capture device and then calls the `save_frames` function to save frames from the video stream. We use `finally` to ensure that we release the video capture device upon exiting the application:

```
def main():
    cap = init_camera()
    try:
        save_frames(cap)
```

```
    finally:
        print('releasing video capture device...')
        cap.release()
```

The full script can be saved as `stream_save.py` on the Pi and then executed.

Listing 10.2 `stream_save.py`: Saving captured video frames to the filesystem

```
#!/usr/bin/env python3
import os
import cv2

IMG_PATH = os.environ['XDG_RUNTIME_DIR'] + '/robo_stream.jpg'
TMP_PATH = os.environ['XDG_RUNTIME_DIR'] + '/robo_stream_tmp.jpg'
FRAME_WIDTH = 320
FRAME_HEIGHT = 240

def save_frames(cap):
    counter = 0
    while True:
        counter += 1
        ret, frame = cap.read()
        assert ret, 'Cannot read frame from camera'
        frame = cv2.flip(frame, -1)
        cv2.imwrite(TMP_PATH, frame)
        os.replace(TMP_PATH, IMG_PATH)
        print('frames:', counter, end='\r', flush=True)

def init_camera():
    cap = cv2.VideoCapture(0)
    assert cap.isOpened(), 'Cannot open camera'
    cap.set(cv2.CAP_PROP_FRAME_WIDTH, FRAME_WIDTH)
    cap.set(cv2.CAP_PROP_FRAME_HEIGHT, FRAME_HEIGHT)
    return cap

def main():
    cap = init_camera()
    try:
        save_frames(cap)
    finally:
        print('releasing video capture device...')
        cap.release()

main()
```

This will continuously capture and save frames from the video stream to the ramdisk. We can execute the following command to list the location of our stream image:

```
$ ls -alh $XDG_RUNTIME_DIR/robo_stream.jpg
-rw-r--r-- 1 robo robo 23K Mar 14 21:12 /run/user/1000/robo_stream.jpg
```

We can see from the output that the image size is 23K, and the file location is `/run/user/1000/robo_stream.jpg`. Each time we open this file in an image viewer, it will show the latest image being captured by the camera.

> **Going deeper: Atomic operations**
>
> Atomic operations are a very powerful and useful feature found in software such as operating systems and databases. They are particularly useful when you have multiple processes accessing the same data at the same time and you want to be sure that you won't face data corruption when reading and writing data. In our case, we want to avoid having one of the streaming applications reading image data that hasn't fully been written to disk. Reading such half-written image data into our applications would cause them to fail. The OSDev website has an excellent page on Atomic operations (https://wiki.osdev.org/Atomic_operation) from an operating system perspective. It is a good resource for further details on the topic.
>
> The Python `os` module documentation (https://docs.python.org/3/library/os.html) covers the `os.replace` function that we use in this section to write the image data to disk as an atomic operation. It mentions that replacing a file on systems such as Linux that follow the Portable Operating System Interface (POSIX) standard will be an atomic operation.
>
> The strategy employed in this chapter of writing to a temporary file and then renaming the file to its final destination is a very common approach used by many applications, such as word processors and web browsers, to ensure data consistency in the final output file.

10.4.2 *Watching the filesystem for changes*

Now that we have our video stream saved to the filesystem, we can read these live video images and display them in different applications. However, to do this, we would need some mechanism where, by polling the filesystem, we could check on a regular basis whether a new image has been made available. One simple and efficient way to do this is to check the modification time of the image file. Whenever it has changed, we know there is a new image available for us. To help the different applications perform this task, we will put the functionality into a library that they can all import and use.

The `sys` module will be used to read command line arguments, and the `time` module will be used to pause between checks for file changes. The `getmtime` function will give us the modification time of the image file:

```
import sys
import time
from os.path import getmtime
```

The `FileWatcher` class receives the `path` to watch and initialize the `last_mtime` attribute to `None` when a new instance is created. Each time the `has_changed` method is called, it gets the current modification time of the file being watched and returns whether this value has changed since it was last checked:

```
class FileWatcher:
    def __init__(self, path):
        self.path = path
        self.last_mtime = None
```

```
    def has_changed(self):
        mtime = getmtime(self.path)
        last_mtime = self.last_mtime
        self.last_mtime = mtime
        return (mtime != last_mtime)
```

The library has a `main` function that can be used to test the `FileWatcher` class. It saves the first command line argument in the `path` variable. Then, it creates a `FileWatcher` instance to watch the specified path. Next, it loops 10 times and checks the file for changes at a rate of 60 frames per second. In each loop, it prints out whether a change was detected:

```
def main():
    path = sys.argv[1]
    print('path:', path)
    watcher = FileWatcher(path)
    for i in range(10):
        print(i, watcher.has_changed())
        time.sleep(1 / 60)
```

The full script can be saved as `watcher.py` on the Pi and then executed.

Listing 10.3 `watcher.py`: Watching a file for changes

```
#!/usr/bin/env python3
import sys
import time
from os.path import getmtime

class FileWatcher:
    def __init__(self, path):
        self.path = path
        self.last_mtime = None

    def has_changed(self):
        mtime = getmtime(self.path)
        last_mtime = self.last_mtime
        self.last_mtime = mtime
        return (mtime != last_mtime)

def main():
    path = sys.argv[1]
    print('path:', path)
    watcher = FileWatcher(path)
    for i in range(10):
        print(i, watcher.has_changed())
        time.sleep(1 / 60)

if __name__ == "__main__":
    main()
```

In one terminal, keep our previous `stream_save.py` running so that it keeps saving the latest frames to the `robo_stream.jpg` file. Then, execute the `watcher.py` script in another terminal and provide it with the stream image to watch. The following session shows the script being executed and the output generated:

```
$ watcher.py $XDG_RUNTIME_DIR/robo_stream.jpg
path: /run/user/1000/robo_stream.jpg
0 True
1 False
2 True
3 False
4 True
5 False
6 True
7 False
8 True
9 False
```

The camera is capturing images at a rate of 30 frames per second, and we are polling the image file for changes at a rate of 60 frames per second. This creates the expected pattern of the file alternating between changed and not changed to exactly match the rate of images being captured from the video stream.

10.4.3 Streaming to a graphical application

With the `watcher` library in place, we can move on and create a graphical application that uses it to watch for changes in the streaming image and display the updated image whenever it changes. The `os` module is imported so that we can access environment variables. The `cv2` module will be used to display images in the application, and the `FileWatcher` will detect changes to the streaming image file:

```
import os
import cv2
from watcher import FileWatcher
```

The `IMG_PATH` variable points to the streaming image file path. `ESC_KEY` has the value for the key code of the Esc key:

```
IMG_PATH = os.environ['XDG_RUNTIME_DIR'] + '/robo_stream.jpg'
ESC_KEY = 27
```

The `main` function creates a `FileWatcher` object in the variable `watcher` and then enters into an event loop. The event loop will keep looping until the Esc key or Q key is pressed. In each loop cycle, the image file is checked for a change by calling the `has_changed` method. If a change is detected, the `imread` function is called to read the new image, and then `imshow` is called to display the image:

```
def main():
    watcher = FileWatcher(IMG_PATH)
    while cv2.waitKey(1) not in [ord('q'), ESC_KEY]:
        if watcher.has_changed():
            img = cv2.imread(IMG_PATH)
            cv2.imshow('preview', img)
```

The full script can be saved as `stream_view.py` on the Pi and then executed.

Listing 10.4 `stream_view.py`: Showing video streaming in a graphical application

```
#!/usr/bin/env python3
import os
import cv2
from watcher import FileWatcher

IMG_PATH = os.environ['XDG_RUNTIME_DIR'] + '/robo_stream.jpg'
ESC_KEY = 27

def main():
    watcher = FileWatcher(IMG_PATH)
    while cv2.waitKey(1) not in [ord('q'), ESC_KEY]:
        if watcher.has_changed():
            img = cv2.imread(IMG_PATH)
            cv2.imshow('preview', img)

main()
```

Make sure to have `stream_save.py` running in another terminal. Now, when you run `stream_view.py`, you can see a live view of the video streaming coming from the camera. Unlike the camera applications in previous chapters, you can start the application multiple times, and each one will stream the video images simultaneously. If you try to do this with the `snapshot.py` application from chapter 8, it will fail because you cannot have more than one application directly capturing frames from the video stream at the same time. With this filesystem-based mechanism of sharing video stream images, we can safely have as many applications as we like accessing and working with live video images. Figure 10.5 shows multiple graphical applications running at the same time and being able to stream the same video stream simultaneously.

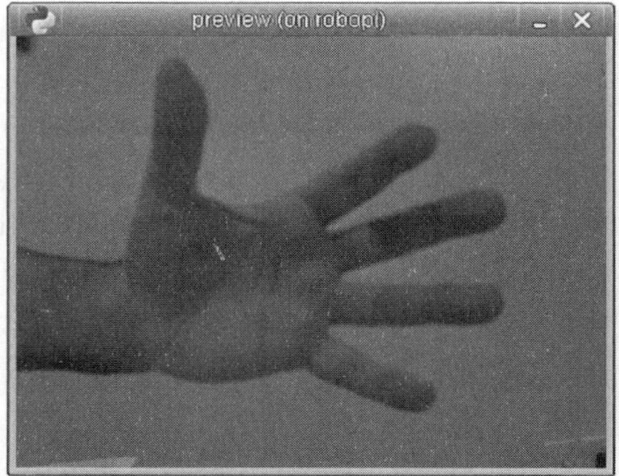

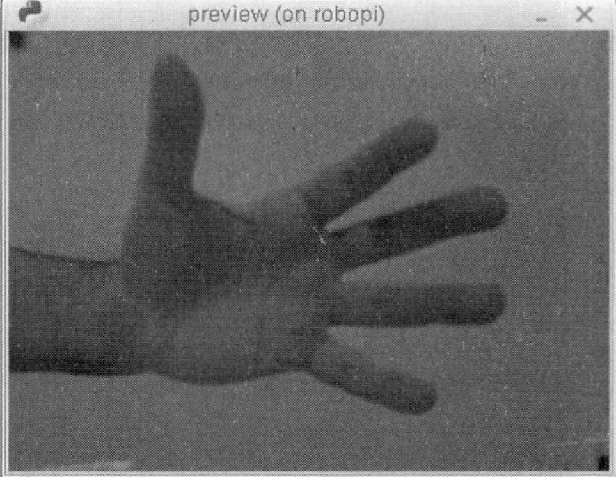

Figure 10.5 Graphical application video streaming: multiple windows can read the video stream.

Since we've got our videos streaming in graphical applications, we can now try and add QR code detection functionality to our video streaming applications.

10.4.4 Detecting QR codes in a video stream

This next application will let us do QR code detection on the live video stream. Any detected QR codes will be marked on the image, and the decoded text will be displayed in the application. This application essentially combines the code and logic from the scripts `detect_qr.py` and `stream_view.py`. We import and use the same three modules from the `stream_view.py` script:

```
import os
import cv2
from watcher import FileWatcher
```

The `IMG_PATH` and `ESC_KEY` variables are taken from `stream_view.py` and serve the same purpose. The `BLUE` and `FONT` variables will be used to set the color and font for drawing in the application. The `decoder` object will perform our QR code decoding:

```
IMG_PATH = os.environ['XDG_RUNTIME_DIR'] + '/robo_stream.jpg'
ESC_KEY = 27
BLUE = (255, 0, 0)
FONT = cv2.FONT_HERSHEY_SIMPLEX
decoder = cv2.QRCodeDetector()
```

The first four lines of the `main` function are identical to the ones in `stream_view.py` and will take care of detecting new images and handling the event loop. Once a new image is detected, the `decode_qrcode` function is called to decode the QR code and draw a box around any detected codes. The `decode_qrcode` and `draw_box` functions are identical to the ones defined in `detect_qr.py`. The last part of the function displays the image by calling `cv2.imshow`:

```
def main():
    watcher = FileWatcher(IMG_PATH)
    while cv2.waitKey(1) not in [ord('q'), ESC_KEY]:
        if watcher.has_changed():
            img = cv2.imread(IMG_PATH)
            decode_qrcode(img)
            cv2.imshow('preview', img)
```

The full script can be saved as `stream_qr.py` on the Pi and then executed.

Listing 10.5 `stream_qr.py`: Detecting QR codes in a streaming video

```
#!/usr/bin/env python3
import os
import cv2
from watcher import FileWatcher

IMG_PATH = os.environ['XDG_RUNTIME_DIR'] + '/robo_stream.jpg'
ESC_KEY = 27
```

```
BLUE = (255, 0, 0)
FONT = cv2.FONT_HERSHEY_SIMPLEX
decoder = cv2.QRCodeDetector()

def draw_box(frame, points, color, thickness):
    points = [(int(x), int(y)) for x, y in points]
    pt1, pt2, pt3, pt4 = points
    cv2.line(frame, pt1, pt2, color, thickness)
    cv2.line(frame, pt2, pt3, color, thickness)
    cv2.line(frame, pt3, pt4, color, thickness)
    cv2.line(frame, pt4, pt1, color, thickness)

def decode_qrcode(frame):
    data, matches, _ = decoder.detectAndDecode(frame)
    if data:
        cv2.putText(frame, f'data: {data}', (30, 30), FONT, 1, BLUE)
        draw_box(frame, matches[0], BLUE, thickness=3)
    return data

def main():
    watcher = FileWatcher(IMG_PATH)
    while cv2.waitKey(1) not in [ord('q'), ESC_KEY]:
        if watcher.has_changed():
            img = cv2.imread(IMG_PATH)
            decode_qrcode(img)
            cv2.imshow('preview', img)

main()
```

Make sure to have `stream_save.py` running in another terminal. Now, when you run `stream_qr.py`, you can see a live view of the video streaming coming from the camera. Any QR codes detected in the image from the video stream will be marked. Figure 10.6 shows the QR code for the QR code used to mark the starting position of the track being detected.

Figure 10.6 Detecting QR codes in live video: detected QR codes are marked on live video.

This script can come very much in handy to test the QR code detection for the printed-out labels. When printing the labels, it is important to not print them out too small, or the camera will not be able to easily detect them. A width and height of 6 cm for the QR codes have been tested and work well. Figure 10.7 shows how the exact dimensions of a QR code label can be set in LibreOffice Writer.

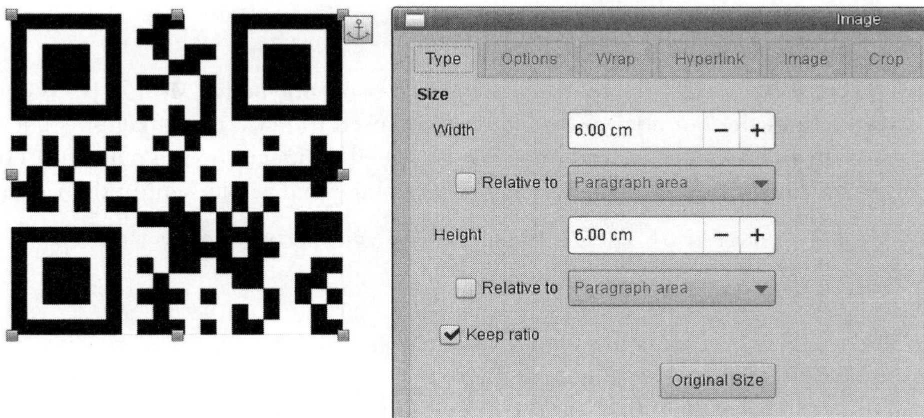

Figure 10.7 QR code label size: it is important to correctly set the size of printed QR codes.

We can now move on to the next challenge of getting the images from the camera to stream into web browsers.

10.4.5 Streaming to a web browser

Streaming the camera video to a format that web browsers can understand opens new and powerful functionality for our robotic web applications, namely, the ability for a web application to get a live video feed and see exactly what the robot sees at that point in time. This will be a new functionality that wasn't available in our previous robot web applications.

The Motion JPEG video format will be used in this application to transmit a continuous stream of video images to the connected web browser. This format is widely used for video streaming and sends a series of JPEG images to the web browser, which are then played back in the web browser like any other video content.

The `os` module will be used to read environment variables, and `FileWatcher` will watch for changes in the image file. The Tornado web framework will be used to create the web application. The `asyncio` is part of the Python standard library and will be used to run the `tornado` main event loop:

```
import os
import asyncio
import tornado.web
from watcher import FileWatcher
```

The `IMG_PATH` variable points to our image file that we will use to check for new images and read them as they are detected. The frequency of polling for changes is defined in `POLL_DELAY` and is set at 60 times per second. It is twice the speed of the camera frame rate, so it should be more than sufficient to detect any new video frames:

```
IMG_PATH = os.environ['XDG_RUNTIME_DIR'] + '/robo_stream.jpg'
POLL_DELAY = 1 / 60
```

The `CONTENT_TYPE` variable stores the HTTP content type for Motion JPEG content. It also defines the boundary value that will be used to mark new images. `BOUNDARY` contains the boundary value and the bytes that need to be sent between images. The `JPEG_HEADER` has the content type for each JPEG image that will be sent in the video stream:

```
CONTENT_TYPE = 'multipart/x-mixed-replace;boundary=image-boundary'
BOUNDARY = b'--image-boundary\r\n'
JPEG_HEADER = b'Content-Type: image/jpeg\r\n\r\n'
```

The `MainHandler` class implements the `get` method, which is called when an incoming HTTP GET request comes to the server and will respond by streaming video content to the browser. It first sets the `Content-Type` of the response to Motion JPEG and then creates a `FileWatcher` object to watch for changes to the stream image file. Next, it enters an infinite loop where, whenever a new image is detected, it is read and sent to the browser with the associated boundary and JPEG HTTP headers. We then call `self.flush` to send the content to the browser. `asyncio.sleep` is called to sleep for the specified polling duration:

```
class MainHandler(tornado.web.RequestHandler):
    async def get(self):
        self.set_header('Content-Type', CONTENT_TYPE)
        watcher = FileWatcher(IMG_PATH)
        while True:
            if watcher.has_changed():
                img_bytes = open(IMG_PATH, 'rb').read()
                self.write(BOUNDARY + JPEG_HEADER + img_bytes)
                self.flush()
            await asyncio.sleep(POLL_DELAY)
```

The `main` function first defines a `tornado.web.Application` object that maps the top-level path to the `MainHandler` class. It then listens on port `9000` for incoming HTTP requests and then calls `shutdown_event.wait()` to wait for a shutdown event:

```
async def main():
    app = tornado.web.Application([('/', MainHandler)])
    app.listen(9000)
    shutdown_event = asyncio.Event()
    await shutdown_event.wait()
```

The full script can be saved as `stream_web.py` on the Pi and then executed.

Listing 10.6 `stream_web.py`: Streaming video to web applications

```
#!/usr/bin/env python3
import os
import asyncio
import tornado.web
from watcher import FileWatcher
```

```
IMG_PATH = os.environ['XDG_RUNTIME_DIR'] + '/robo_stream.jpg'
POLL_DELAY = 1 / 60
CONTENT_TYPE = 'multipart/x-mixed-replace;boundary=image-boundary'
BOUNDARY = b'--image-boundary\r\n'
JPEG_HEADER = b'Content-Type: image/jpeg\r\n\r\n'

class MainHandler(tornado.web.RequestHandler):
    async def get(self):
        self.set_header('Content-Type', CONTENT_TYPE)
        watcher = FileWatcher(IMG_PATH)
        while True:
            if watcher.has_changed():
                img_bytes = open(IMG_PATH, 'rb').read()
                self.write(BOUNDARY + JPEG_HEADER + img_bytes)
                self.flush()
            await asyncio.sleep(POLL_DELAY)

async def main():
    app = tornado.web.Application([('/', MainHandler)])
    app.listen(9000)
    shutdown_event = asyncio.Event()
    await shutdown_event.wait()

asyncio.run(main())
```

In one terminal, keep our previous `stream_save.py` running so that it keeps saving the latest frames to the `robo_stream.jpg` file. Then execute the `stream_web.py` script in another terminal. You can access the web application by visiting the address `http://robopi:9000` from a computer on your network. You can also access the web app by replacing RoboPi in the URL with the IP address of your robot. When accessing the web app from a mobile device, using the IP address will be an easier option. Figure 10.8 shows what the video stream looks like when accessed on a mobile device. In this example, the live video stream was viewed on an Android mobile device over a Wi-Fi network.

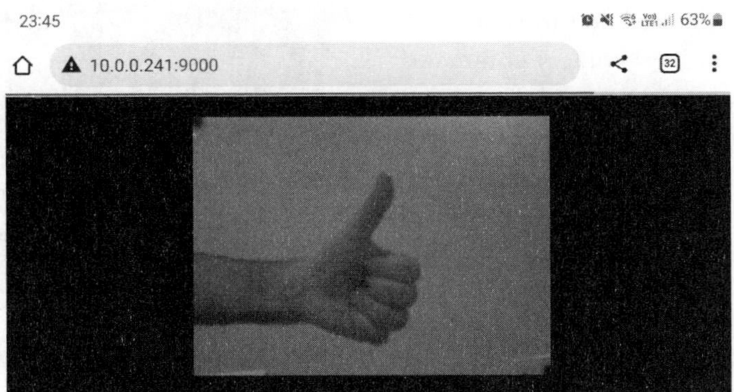

Figure 10.8 Web application video streaming: the image shows streaming over the web to a mobile device.

Compared to the graphical application, the web-based approach offers greater flexibility, as any modern web browsers on any mobile or desktop computer can be used to access the video stream. Once again, there is the added benefit that many computers and devices can access and view the video stream simultaneously without any problems.

10.5 Moving the robot to a target QR code

We can now take on the final challenge in this chapter of driving the robot along the track until it finds a specific QR code. We need to create a Python application that meets the following requirements:

- The name of the target QR code should be provided as the first command line argument.
- The robot should be driven in the forward direction while continually scanning for QR codes.
- The robot should stop when the target QR code is first detected.

We can place many objects with QR codes on them along the robot track and use this technique to ask the robot to go to one of these specific locations.

10.5.1 Find the QR code

As the robot moves along the track, it will keep checking the video stream for any detected QR codes. If it finds a QR code, it will compare its value to the target that we are looking for. If it finds a match, we stop the robot. For safety, we will also provide a maximum number of moves the robot can make so that it doesn't crash into the end of the track. Figure 10.9 shows the camera mounted on the side of the robot right above the wheel so that it can capture QR codes of the objects it drives by.

Figure 10.9 Side camera: the camera is mounted on the side of the robot above the wheel.

The `os` module will be used to read environment variables, and `sys` will get command line arguments. We will use `cv2` to do the QR detection and `motor` to move the robot motors:

```
import os
import sys
import cv2
import motor
```

The `MAX_MOVES` variable sets a limit of 20 moves before the robot gives up on finding its target. `IMG_PATH` points to the video stream image, and `decoder` will be used to decode the QR codes:

```
MAX_MOVES = 20
IMG_PATH = os.environ['XDG_RUNTIME_DIR'] + '/robo_stream.jpg'
decoder = cv2.QRCodeDetector()
```

The `decode_qr` function reads the latest image from the video stream and attempts to decode any QR codes found in the image. The decoded data is then returned:

```
def decode_qr():
    img = cv2.imread(IMG_PATH)
    data, points, _ = decoder.detectAndDecode(img)
    return data
```

The `goto` function loops the number of times specified in `MAX_MOVES`. In each loop, it moves the robot forward at the lowest speed for 0.1 seconds. It then decodes QR code data from the latest video image and prints out its progress so far, as well as whatever data it has just decoded. If the decoded data matches the value of `target`, we return the `True` value to indicate a successful search. If we have exceeded `MAX_MOVES`. then we return `False` to indicate the search for `target` was unsuccessful:

```
def goto(target):
    for i in range(MAX_MOVES):
        motor.forward(speed=1, duration=0.1)
        data = decode_qr()
        print(f'searching {i + 1}/{MAX_MOVES}, data: {data}')
        if data == target:
            return True
    return False
```

The `main` function gets the value of `target` from the first command line argument. It prints out the value and then calls `goto` with `target`. Finally, the result of the search is saved and printed out:

```
def main():
    target = sys.argv[1]
    print('target:', repr(target))
    found = goto(target)
    print('found status:', found)
```

The full script can be saved as `goto_qr.py` on the Pi and then executed.

Listing 10.7 `goto_qr.py`: Searching and going to the target QR code

```
#!/usr/bin/env python3
import os
import sys
import cv2
import motor

MAX_MOVES = 20
IMG_PATH = os.environ['XDG_RUNTIME_DIR'] + '/robo_stream.jpg'
```

```
decoder = cv2.QRCodeDetector()

def decode_qr():
    img = cv2.imread(IMG_PATH)
    data, points, _ = decoder.detectAndDecode(img)
    return data

def goto(target):
    for i in range(MAX_MOVES):
        motor.forward(speed=1, duration=0.1)
        data = decode_qr()
        print(f'searching {i + 1}/{MAX_MOVES}, data: {data}')
        if data == target:
            return True
    return False

def main():
    target = sys.argv[1]
    print('target:', repr(target))
    found = goto(target)
    print('found status:', found)

main()
```

Make sure to keep `stream_save.py` running in another terminal so that it keeps saving the latest frames to the filesystem. Then, execute the `goto_qr.py` script in another terminal. You can also watch what the robot is seeing by using either `stream_view.py` or `stream_web.py`. The following session shows the script being executed and the output it generated:

```
$ goto_qr.py start
target: 'start'
searching 1/20, data:
searching 2/20, data:
searching 3/20, data:
searching 4/20, data: start
found status: True
```

We asked the robot to search for a target called `start`. The robot made four movements before successfully going to the marker with the `start` QR code. The search then ended and was reported as a success. Let's see what happens when we search for a target further down the track:

```
$ goto_qr.py end
target: 'end'
searching 1/20, data:
searching 2/20, data:
searching 3/20, data:
searching 4/20, data: start
searching 5/20, data: start
searching 6/20, data: start
searching 7/20, data: start
searching 8/20, data: start
searching 9/20, data:
```

```
searching 10/20, data:
searching 11/20, data:
searching 12/20, data: end
found status: True
```

We can see that the robot encountered the `start` marker again during movement number `4`. As it kept searching, it finally got to the target `end` at movement number `12`. Like before, it then returned with a return value indicating a successful search. Let's try asking the robot to find a target that doesn't exist and see what happens:

```
$ goto_qr.py never_find_me
target: 'never_find_me'
searching 1/20, data:
searching 2/20, data:
searching 3/20, data:
searching 4/20, data:
searching 5/20, data: start
searching 6/20, data: start
searching 7/20, data: start
searching 8/20, data: start
searching 9/20, data: start
searching 10/20, data:
searching 11/20, data:
searching 12/20, data:
searching 13/20, data:
searching 14/20, data: end
searching 15/20, data: end
searching 16/20, data: end
searching 17/20, data: end
searching 18/20, data:
searching 19/20, data:
searching 20/20, data:
found status: False
```

The robot has exceeded the maximum number of permitted movements without finding its target. It returned with a `False` value to indicate the search was unsuccessful. This covers the scenario of searching for an object that does not exist.

Robots in the real world: Warehouse robots

Robots have been increasingly used in warehouses to retrieve items that are to be shipped to customers. The Smart Warehouse (http://mng.bz/ored) article by Supply Chain Today shows different types of automation that Amazon uses in its warehouse.

The mobile robots used in the warehouse to move items around different locations employ an interesting navigation mechanism. They have cameras on the bottom that read QR codes on the floor. The warehouse floor is filled with QR codes in a grid pattern that the robots can read to find out exactly where they are in the warehouse. This approach is very similar to how the robot in this chapter reads QR codes in its environment to navigate to a specific location.

Summary

- The OpenCV computer vision library is used to detect QR codes in images, as well as to read the data encoded in the QR code itself.
- The robot will use the DC motors to move back and forth along a set track.
- The more data you put in a QR code, the larger the symbol version becomes, which in turn generates a denser QR code.
- Detected QR codes have four points relating to the four corners of a quadrangle.
- A ramdisk is used to stream video images, as this will not create additional disk workload.
- Changes to the streaming image are checked by polling the filesystem on a regular basis.
- The Motion JPEG video format is used in web video streaming applications to transmit a continuous stream of video images to web browsers.

11 Building a snack-pushing robot

This chapter covers

- Reading a list of QR codes and icons from a CSV file
- Locating and pushing selected objects
- Creating a user interface for video streaming and snack selection
- Building a snack-pushing robot

This chapter aims to build a snack-pushing robot that can be controlled by a web-based Python application. The chapter starts by reading a list of snacks from a CSV file. Each snack will have a QR code and icon assigned. The QR code will be used to help the robot find the snack. The snack icon will be displayed with the code in the web application. Then, we take on the challenge of moving the robot to the selected snack and positioning it in an ideal position to push the snack off the ledge and into the hands of a hungry snack-eating human. The robot will then return to the starting position and wait for another snack request. In the final part of the chapter, we create a user interface and a web application that shows a live video stream from the robot camera and provides a list of available snacks. Select the snack, and watch the robot push it off the edge of the table.

This application can be used as a launching point to create many different types of applications that can be controlled from a mobile device and have the robot seek and fetch different items from its environment. Robots that can drive around and use an arm to interact with their surroundings are quite versatile in their use.

11.1 Hardware stack

Figure 11.1 shows the hardware stack, with the specific components used in this chapter highlighted. The robot will use the DC motors to move along the track in search of a specific target QR code. Images will be captured from the camera, and QR code detection will be applied on these images until a match is found. Next, the robot's motors will stop. The robot will then use the motors to position the servo arm in an ideal position to push the detected snack. At this stage, the arm attached to the servo motor will be moved up and down to push the snack off the counter. For tips on how to position the snacks so that the robot can detect and push them with ease, refer to appendix C.

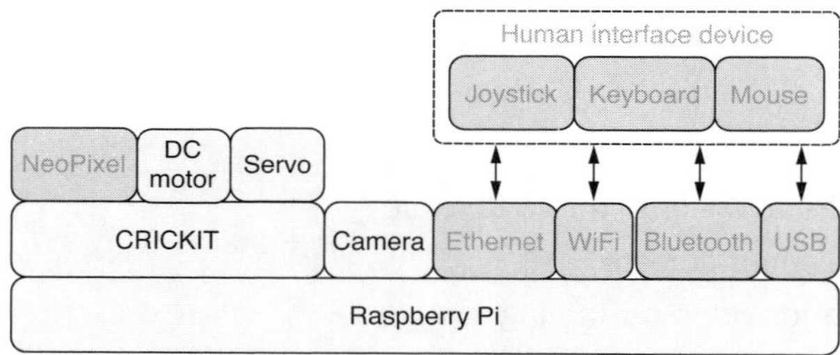

Figure 11.1 Hardware stack: the servo motors will be used to push the snacks.

11.2 Software stack

Details of the specific software used in this chapter are shown in figure 11.2. We first read the list of snacks from a CSV file using the `csv` library. Each snack has an emoji icon that is converted using the `emoji` library. Next, we create the `pusher_qr` library that will detect QR codes using OpenCV and push the snacks using the servo motors. We will employ the Tornado web framework to create the `pusher_web` application to allow users to control the robot from their mobile devices.

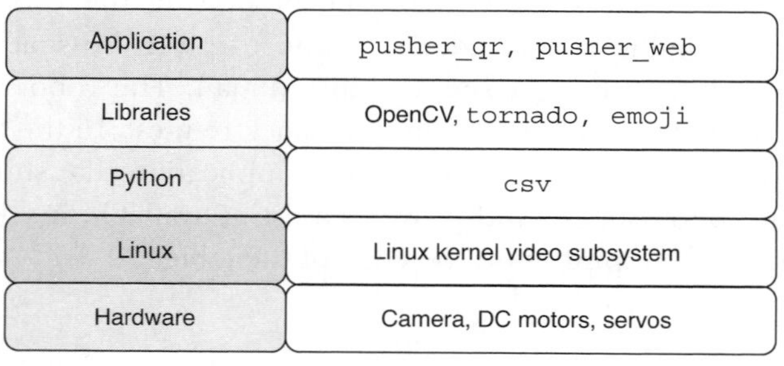

Figure 11.2 Software stack: The `emoji` library will be used to convert emoji short codes for the icons.

11.3 Finding and pushing snacks

The first step will be to create a library to locate and push snacks based on their QR code. We need to create a Python library that meets the following requirements:

- It should read the list of snacks from a CSV file and convert any emoji short codes found.
- The library should have a function that will swing the servo arm up and down.
- The library should have a function that looks for a matching QR code and swings the arm when the code is found.

This library will provide us with the core functionality for our robot. Figure 11.3 shows a side view of the robot with the camera used to detect QR codes and the servo arm to push snacks.

Figure 11.3 Snack-pushing robot: the servo motor is used to push snacks.

Once the library is in place, we will be able to develop a web application to call different robot functions as needed.

11.3.1 Reading the list of snacks

The first step is to install the emoji Python package. This module will let us convert emoji short codes to Unicode characters. We will use this package to create the icons in the application for each snack. Run the following command to install the package:

```
$ ~/pyenv/bin/pip install emoji
```

Now that we have everything we need installed, let's open a read–evaluate–print loop (REPL) session and get to it. First, we will tackle the task of reading and parsing the snack list from a CSV file. We import the `DictReader` object to read CSV data as a list of dictionaries. We then import the function `pprint` to do a pretty print of our data structures:

```
>>> from csv import DictReader
>>> from pprint import pprint
```

The CSV file should be saved as `items.csv` on the Pi with the contents presented in listing 11.1.

Listing 11.1 `items.csv`: List of snack QR codes and icons

```
code,icon
grapes,:grapes:
carrots,:carrot:
candy,:candy:
lollipop,:lollipop:
```

The first line of the file contains the field names. The `code` field stores the value for the QR code, and the `icon` field stores the value of the icon as an emoji short code. The first step will be to read the lines from the CSV file with the following line of code:

```
>>> lines = list(open('items.csv'))
```

We can now have a peek at what is in `lines`. It contains a list of strings. Each string is one line in the file:

```
>>> pprint(lines)
['code,icon\n',
 'grapes,:grapes:\n',
 'carrots,:carrot:\n',
 'candy,:candy:\n',
 'lollipop,:lollipop:\n']
```

We use `DictReader` to parse the lines and return a list of dictionaries:

```
>>> items = list(DictReader(lines))
```

We can now pretty print `items` to get a better look at what is inside:

```
>>> pprint(items)
[{'code': 'grapes', 'icon': ':grapes:'},
 {'code': 'carrots', 'icon': ':carrot:'},
 {'code': 'candy', 'icon': ':candy:'},
 {'code': 'lollipop', 'icon': ':lollipop:'}]
```

We can grab the first item from the list and inspect the `code` and `icon` for that item:

```
>>> items[0]
{'code': 'grapes', 'icon': ':grapes:'}
>>> items[0]['code']
'grapes'
>>> items[0]['icon']
':grapes:'
```

Now, let's move on to converting emoji short codes. We will import the `emojize` function to convert the short codes and use `pathlib` to save a test HTML file to disk:

```
>>> from emoji import emojize
>>> from pathlib import Path
```

The emoji package page (https://pypi.org/project/emoji/) provides good documentation on using the module and links to the Unicode consortium pages that have a list of the official emoji short codes. Let's convert some text by calling the `emojize` function:

```
>>> text = emojize('Have some :pie: with your pi!')
```

We want to see what the end result will look like in our web browser, so let's add the text to some HTML and save it to a file:

```
>>> html = '<!DOCTYPE html><title>_</title>' + text
>>> Path('pie.html').write_text(html)
```

Now, when we open the `pie.html` file in our web browser, we will be able to see what these emoji icons will look like. Figure 11.4 shows this HTML file as it would be displayed in a web browser.

Now that we have read our list of snacks and figured out how to create some emoji icons, let's move on to finding and pushing snacks with the robot.

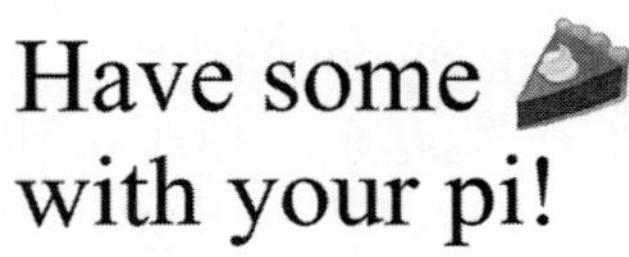

Figure 11.4 Emoji short codes: the emoji short code for pie was converted to Unicode.

11.3.2 Pushing snacks

We will create a library with several functions to help us locate and push snacks. We import the `dirname` function to get the path of our Python file and `csv` to parse our CSV snacks list. Then, we import `emojize` to help with the emoji icons and `crickit` to control the servo motors:

```
from os.path import dirname
from csv import DictReader
from emoji import emojize
from adafruit_crickit import crickit
```

Next, we import `motor` to handle the forward and backward movements on our DC motors. The `os` module will access environment variables, and `time` will be used to pause between the servo arm movements. The `cv2` module will help perform QR code detection:

```
import motor
import os
import time
import cv2
```

The constant ITEMS_FILE points to our CSV file and IMG_PATH to our streaming image file. We limit the movement of the robot with MAX_MOVES and define the servo angles to move the servo arm up and down in SERVO_ANGLES. The decoder object will decode our QR codes:

```
ITEMS_FILE = dirname(__file__) + '/items.csv'
IMG_PATH = os.environ['XDG_RUNTIME_DIR'] + '/robo_stream.jpg'
MAX_MOVES = 20
SERVO_ANGLES = dict(up=70, down=180)
decoder = cv2.QRCodeDetector()
```

The get_items function opens our CSV file and converts all the emoji short codes for every line in the file. Next, we call DictReader to parse the CSV content and return a list of dictionaries:

```
def get_items():
    lines = [emojize(i) for i in open(ITEMS_FILE)]
    return list(DictReader(lines))
```

Our reliable decode_qr function will do the job of decoding any QR codes we encounter:

```
def decode_qr():
    img = cv2.imread(IMG_PATH)
    data, points, _ = decoder.detectAndDecode(img)
    return data
```

The goto function moves the robot in the provided direction looking for a QR-code-matching target. We use direction to look up our movement function and save it in motor_func. Then, we loop through moving our robot in the desired direction and calling decode_qr to check whether we have encountered any QR codes. If we find a code-matching target, we return with a True value. Otherwise, if we are moving forward and reach the end of the track, we return with False. Likewise, if we have exceeded MAX_MOVES movement attempts, we return with False:

```
def goto(target, direction):
    motor_func = getattr(motor, direction)
    for i in range(MAX_MOVES):
        motor_func(speed=1, duration=0.1)
        data = decode_qr()
        if data == target:
            return True
        if data == 'end' and direction == 'forward':
            return False
    return False
```

We use swing_arm to swing our servo arm up and knock the snacks over. We pause for half a second and swing the arm back down to its original position. The same servo motor is used to move the arm to the up and down positions. Figure 11.5 shows the arm in the down position, which is used when driving along the track. Figure 11.6 shows the arm in the up position, which is used to knock over the snacks:

```
def swing_arm():
    crickit.servo_2.angle = SERVO_ANGLES['up']
    time.sleep(0.5)
    crickit.servo_2.angle = SERVO_ANGLES['down']
    time.sleep(0.5)
```

The `push_item` function is used to drive the robot forward in search of a QR-code-matching `code`. If found, we move the robot backward to position the servo arm in the center of our snack, and then we swing the arm by calling `swing_arm`. Finally, we call `goto` to drive the robot back to the starting position:

```
def push_item(code):
    found = goto(code, 'forward')
    if found:
        motor.backward(speed=1, duration=0.3)
        swing_arm()
    goto('start', 'backward')
```

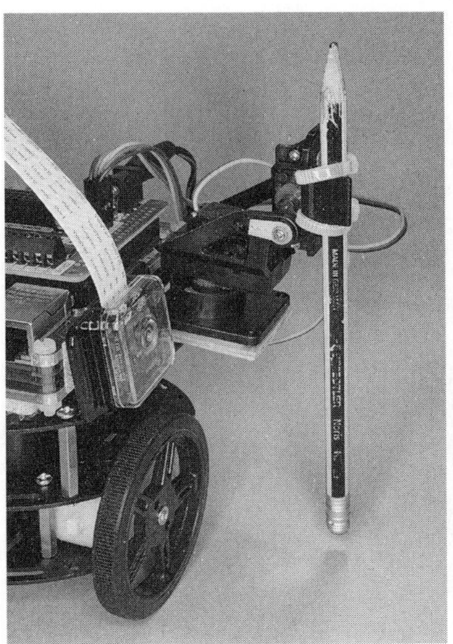

Figure 11.5 Arm down: the arm is kept in the down position when the robot is moving on the track.

Figure 11.6 Arm up: the arm is placed in the up position to knock over snacks.

The full script can be saved as `pusher_qr.py` on the Pi and then executed.

Listing 11.2 `pusher_qr.py`: Library to detect and push matching snacks

```
#!/usr/bin/env python3
from os.path import dirname
from csv import DictReader
```

```
from emoji import emojize
from adafruit_crickit import crickit
import motor
import os
import time
import cv2

ITEMS_FILE = dirname(__file__) + '/items.csv'
IMG_PATH = os.environ['XDG_RUNTIME_DIR'] + '/robo_stream.jpg'
MAX_MOVES = 20
SERVO_ANGLES = dict(up=70, down=180)
decoder = cv2.QRCodeDetector()

def get_items():
    lines = [emojize(i) for i in open(ITEMS_FILE)]
    return list(DictReader(lines))

def decode_qr():
    img = cv2.imread(IMG_PATH)
    data, points, _ = decoder.detectAndDecode(img)
    return data

def goto(target, direction):
    motor_func = getattr(motor, direction)
    for i in range(MAX_MOVES):
        motor_func(speed=1, duration=0.1)
        data = decode_qr()
        if data == target:
            return True
        if data == 'end' and direction == 'forward':
            return False
    return False

def swing_arm():
    crickit.servo_2.angle = SERVO_ANGLES['up']
    time.sleep(0.5)
    crickit.servo_2.angle = SERVO_ANGLES['down']
    time.sleep(0.5)

def push_item(code):
    found = goto(code, 'forward')
    if found:
        motor.backward(speed=1, duration=0.3)
        swing_arm()
    goto('start', 'backward')
```

We can take this library for a test drive now. Just like we have done in the previous chapter, make sure to have `stream_save.py` running in another terminal. Place the robot at the starting position of the track pointing at the starting QR code. We can try out the library in a REPL session. First, we import the `pusher_qr` library:

```
>>> import pusher_qr
```

We call the `decode_qr` function, and it returns the code for our starting position as `start`:

```
>>> pusher_qr.decode_qr()
'start'
```

We can now ask the robot to go to the end of the track with the following function call:

```
>>> pusher_qr.goto('end', 'forward')
True
```

The function returned `True`, which means it successfully reached the end position. We can call `decode_qr` to confirm this. The function returns the `end` value:

```
>>> pusher_qr.decode_qr()
'end'
```

Next, we return to the start position:

```
>>> pusher_qr.goto('start', 'backward')
True
```

Now, let us push the snack with code `candy` by calling the `push_item` function. The robot will move to the snack with QR code `candy`, push it with the servo arm, and then return to the starting position:

```
>>> pusher_qr.push_item('candy')
```

Like before, we can confirm the robot is at the starting position by calling `decode_qr`:

```
>>> pusher_qr.decode_qr()
'start'
```

This session was a good way to take the library and the robot for a test drive before putting our web application in place as a front end to control the robot.

Robots in the real world: Pick-and-place robots

One very popular category of robots is pick-and-place robots. They are often used in manufacturing settings where produced items need to be packed so that they can be shipped. The robot in this chapter has the ability to locate specific items and push them off a counter. Imagine pushing the items onto a conveyor belt so that they can be taken to another part of a factory for further processing.

Some of the benefits of rolling out pick-and-place robots are the increased speed and reliability compared to manual picking and placing. They come in many shapes and sizes, depending on the types of items they need to pick up and their characteristics. The 6 River Systems site (https://6river.com/what-is-a-pick-and-place-robot) covers the topic of pick-and-place robots very well and is a good place to learn more about different types of robots that are commonly used and their applications.

11.4 Creating the snack-pushing application

We can now dive into creating our web application to control our snack-pushing robot. We need to create a Python application that meets the following requirements:

- It should show a list of snacks as buttons for selection to be made.
- Once a snack is selected, the robot should drive to the snack and push it. Then it should return to the starting position.
- A live video stream of the robot camera should be included in the application.

There are a number of challenges ahead of us, so let's break up the problem a bit. First, we will tackle listing and selecting items. Then, we will focus on how to use stylesheets to better control the layout and design of our user interface. Finally, we will add the live video stream to the application.

11.4.1 Selecting snacks with the application

We will first focus on reading the list of snacks and presenting them as a series of buttons. Clicking on one of these snack buttons will then have our robot drive to the snack and do its magic.

As we have done before, we import a number of functions and objects from the Tornado web framework to help us create our web application. These are all the same functions and objects we have used in previous chapters:

```
from tornado.ioloop import IOLoop
from tornado.web import RequestHandler, Application
from tornado.log import enable_pretty_logging
```

We then import from the `os` module to obtain directory names and environment variables. We import `get_items` and `push_item` to list the available and push selected items:

```
from os.path import dirname
import os
from pusher_qr import get_items, push_item
```

We save the settings for our application in `SETTINGS`. We use `static_path` so that we can serve static content like stylesheets:

```
SETTINGS = dict(
    debug=bool(os.environ.get('ROBO_DEBUG')),
    template_path=dirname(__file__) + '/templates',
    static_path=dirname(__file__) + '/static',
)
```

The `MainHandler` object will handle incoming requests. For `GET` requests, we save the list of snacks and pass it in `items` to the template to render. When the index page is accessed, `name` will be blank, so we set it to the value `index`. Otherwise, the name of the page being accessed is directly mapped to the template name. When the snack selection form is submitted, the `post` method will handle the request by calling push_item to push the item and then calling redirect to take the browsers to the page listing all the items:

```
class MainHandler(RequestHandler):
    def get(self, name):
        name = name or 'index'
        self.render(f'{name}.html', items=get_items())

    def post(self, code):
        push_item(code)
        self.redirect('items')
```

The final step is like the one we have already seen. We enable pretty logging and then create our application objection and have it listen on port 8888 for incoming requests:

```
enable_pretty_logging()
app = Application([('/([a-z_]*)', MainHandler)], **SETTINGS)
app.listen(8888)
IOLoop.current().start()
```

The full script can be saved as pusher_web.py on the Pi and then executed.

Listing 11.3 pusher_web.py: Handling requests for the snack-pusher application

```
#!/usr/bin/env python3
from tornado.ioloop import IOLoop
from tornado.web import RequestHandler, Application
from tornado.log import enable_pretty_logging
from os.path import dirname
import os
from pusher_qr import get_items, push_item

SETTINGS = dict(
    debug=bool(os.environ.get('ROBO_DEBUG')),
    template_path=dirname(__file__) + '/templates',
    static_path=dirname(__file__) + '/static',
)

class MainHandler(RequestHandler):
    def get(self, name):
        name = name or 'index'
        self.render(f'{name}.html', items=get_items())

    def post(self, code):
        push_item(code)
        self.redirect('items')

enable_pretty_logging()
app = Application([('/([a-z_]*)', MainHandler)], **SETTINGS)
app.listen(8888)
IOLoop.current().start()
```

Before running this script, we will need to at least create one HTML template to be served to the web browser. Ultimately, the application will have a template to display the index page and one to display the list of snacks. We will tackle the list of snacks template first. Let's have a look at the contents of this HTML template.

We start with the header portion of the HTML document. Here, we set the title of the page and use the `meta` tag to ensure the page renders well on mobile devices. Like before, we set a blank icon for the page. We then point to a stylesheet called `style.css` that will be part of our static content. We use the Tornado `static_url` function to generate the URL for this static content:

```
<!DOCTYPE HTML>
<html lang="en">
<head>
  <title>Snack Pusher</title>
  <meta name="viewport" content="width=device-width">
  <link rel="icon" href="data:,">
  <link rel="stylesheet" href="{{ static_url('style.css') }}">
</head>
```

We now move on to the body of the document, which contains a form to be submitted using the POST method. We loop through each snack in the `items` variable. For each snack, we output a button with action defined by `code`. The text of the button will show both the value of `icon` and `code`:

```
<body>
<form method="post">
  {% for item in items %}
    <button formaction="{{ item['code'] }}">
      {{ item['icon'] }}<br>
      {{ item['code'] }}
    </button>
  {% end %}
</form>
</body>
</html>
```

The full template can be saved as `items.html` in the templates folder of the application.

Listing 11.4 `items.html`: HTML template showing the list of available items

```
<!DOCTYPE HTML>
<html lang="en">
<head>
  <title>Snack Pusher</title>
  <meta name="viewport" content="width=device-width">
  <link rel="icon" href="data:,">
  <link rel="stylesheet" href="{{ static_url('style.css') }}">
</head>
<body>
<form method="post">
```

```
  {% for item in items %}
    <button formaction="{{ item['code'] }}">
      {{ item['icon'] }}<br>
      {{ item['code'] }}
    </button>
  {% end %}
</form>
</body>
</html>
```

Now, we have enough of our application in place to run it and start testing parts of its functionality. Like before, make sure to have `stream_save.py` running in another terminal. Now go ahead and run our new `pusher_web.py` application. We can access the web application using a web browser on any computer or mobile device on the same network as the robot. Access the web application by visiting the address http://robopi:8888/items or by replacing `robopi` in the URL with the IP address of your robot. Figure 11.7 shows what this part of the application looks like. It has a list of our four defined snacks in our CSV file. Each snack has its icon and name shown. Press one of these buttons, and the robot will drive along, find the selected snack, and push it off of the counter.

We have a good level of functionality in place. Now, let's add some style to this app using stylesheets.

Figure 11.7 Item listing: the list of snacks is shown in the application.

11.4.2 Styling the web application

We will create one stylesheet to style both our pages. We have some style elements common to both pages, so it makes sense to keep all the styles in one stylesheet.

We first style the content in the main body, links, and buttons. We set the font to use, center the text, and remove the link underlining by setting `text-decoration` to `none`:

```
body, a, button {
  font-family: Verdana, Arial, sans-serif;
  text-align: center;
  text-decoration: none;
}
```

We will double the font size on the buttons and add a healthy amount of margin and padding to make them bigger and easier to press on mobile devices. We set them all to the same width of 140 px so that they can have a uniform size:

```
button {
  font-size: 200%;
  padding: 10px;
  margin: 10px;
  width: 140px;
}
```

In the next section, we will add a template for the index page. That page has an `iframe` that we would like to style. We make the `iframe` take up the full screen width and have a height of 300 px. We also remove the border so that it fits the look of the page more naturally:

```
iframe {
  width:100%;
  height:300px;
  border:none;
}
```

The stylesheet can be saved as `style.css` in the static content folder of the application.

Listing 11.5 `style.css`: Applying a stylesheet to the HTML templates

```
body, a, button {
  font-family: Verdana, Arial, sans-serif;
  text-align: center;
  text-decoration: none;
}

button {
  font-size: 200%;
  padding: 10px;
  margin: 10px;
  width: 140px;
}

iframe {
  width:100%;
  height:300px;
  border:none;
}
```

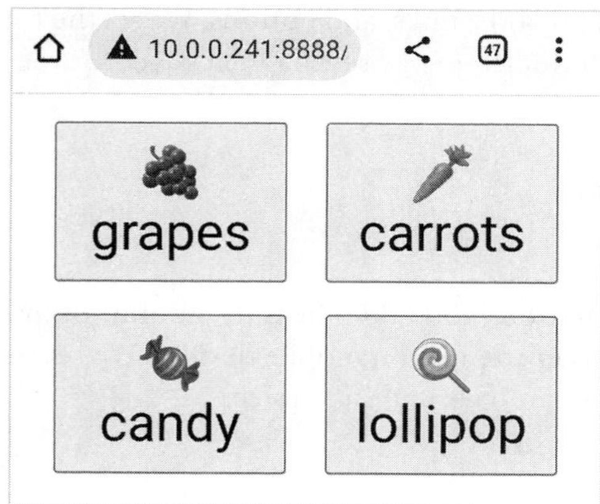

Figure 11.8 Styled buttons: the buttons have been styled with a bigger font and more padding.

Now we can start our `pusher_web.py` application again to have a peek at our application. Visit the same URL to see how the page has changed. Figure 11.8 shows the new look for the page once the styles are applied. The buttons will appear much larger now and will be much easier to press on mobile devices with smaller screens.

With the stylesheets out of the way, we can tackle the final part of the application. The index page will combine the snack list with the live video stream.

11.4.3 Adding the live video stream to the application

Adding the live video stream to web applications is relatively simple. We just put the URL of the video streaming service into an `img` tag. The problem is that each time we select a snack by pressing a button, the web browser will submit the whole page. This will refresh the whole page and make us miss the most exciting part of the video stream, which is the robot rolling down its track in search of our snack. We can address this problem by putting our list of snacks in their own `iframe`. In that way, our video stream playback will never be interrupted, no matter how many snacks we select. We can now have a look at the template for the main index page.

We have our usual tags in the header to set the language and title of the page. All the tags and values in the header are identical to what we have used in `items.html`:

```
<!DOCTYPE HTML>
<html lang="en">
<head>
  <title>Snack Pusher</title>
  <meta name="viewport" content="width=device-width">
  <link rel="icon" href="data:,">
  <link rel="stylesheet" href="{{ static_url('style.css') }}">
</head>
```

In the body, we put a header with the page title at the top of the page. Then, we place our live video stream right after the header. We make sure to use the `host_name` value in the request so that the application works correctly regardless of whether you accessed it by the name of the host or IP address. Next, we load our page with the list of snacks in an `iframe`, right beneath the live video stream:

```
<body>
<h1><a href='/'>Snack Pusher</a></h1>
<img src="http://{{request.host_name}}:9000/" alt="camera"><br>
<iframe src="/items"></iframe>
</body>
</html>
```

The full template can be saved as `index.html` in the templates folder of the application.

Listing 11.6 `index.html`: Template showing the live video stream and snack list

```
<!DOCTYPE HTML>
<html lang="en">
<head>
  <title>Snack Pusher</title>
  <meta name="viewport" content="width=device-width">
  <link rel="icon" href="data:,">
  <link rel="stylesheet" href="{{ static_url('style.css') }}">
</head>
<body>
<h1><a href='/'>Snack Pusher</a></h1>
<img src="http://{{request.host_name}}:9000/" alt="camera"><br>
```

```
<iframe src="/items"></iframe>
</body>
</html>
```

All the pieces of our application are now in place, and we can try out the final version. In addition to having `stream_save.py` running, make sure `stream_web.py` is also running so that the live video stream can be served to the application. Run the `pusher_web.py` script so that we can try our application. Access the web application by visiting the address http://robopi:8888/ or by the IP address of the robot. Figure 11.9 shows what the full application will look like. We can now see the live video stream and make our snack selections in the same application. We can select any of the snacks, and as the request is processed by the robot, the video stream will keep playing uninterrupted.

Figure 11.9 Final application: the live video stream is shown while snacks are selected.

> **Going deeper: Video streams in applications**
>
> This is the first time that we embed a live video stream into a web application, which is a very powerful feature of the HTML language. We can expand this functionality in many ways.
>
> We can create user interfaces with multiple video streams being displayed simultaneously. This is particularly useful for robots that have multiple cameras on board, with one facing forward and another facing backward. By displaying all the video feeds from the cameras at the same time, we can obtain a full view of the robot's environment as it drives around.
>
> Another useful feature that we could add is the ability to continuously record and play back video streams. This could be done by creating a stream-archiving application that would save each new frame from the video stream into time-stamped filenames. Then we would add the option in the user interface to rewind back in time. If we left the video stream to be continually saved, the disk would eventually fill up, and we would run out of storage. We could address this by implementing a data-retention policy where we would only retain video stream data for the past 24 hours. Any older data would be removed automatically, and then the application could maintain its

record and playback features and not run out of disk space. Alternatively, we could sync the old video data to a centralized video archive server on the network. This is another common strategy for dealing with the limited local storage on robots.

This last project has combined many different technologies to create a robot that can interact with its environment in powerful ways, and its functionality can be extended. We could add a function where the robot drives around and automatically does a stock check on the available items. Instead of pushing the items, we could create a pick-and-place robot that grabs items and places them in another location, similar to what a warehouse robot would do. There are many ways we could improve our robots, and we are only limited by our imagination.

And there you have it. This is the end of a long fun journey bringing together a lot of different hardware and software components to make a robot we can control with a mobile device over a wireless network. With mobile in hand, select as many snacks as you like, and enjoy the robot whizzing around, searching and tossing whatever snacks your heart desires.

Summary

- The snack-pushing robot is controlled by a web-based Python application.
- The servo motor is used to move the servo arm up and down and push the snacks.
- The emoji Python package is used to convert emoji short codes to Unicode characters. These are used as icons for the snacks in the application.
- The list of snacks is read from a CSV file that contains the QR code and icon for each snack.
- A single stylesheet is used to style both template pages in the application. This is done because there are common style elements in both pages, so it is easier to have all styles in one stylesheet.
- An `iframe` is used to help us play the live video stream and submit snack selections in the same application without any interruption to video playback.

appendix A Hardware purchasing guide

This appendix covers different hardware components necessary to build the robots covered in this book. It provides details on the specific models of products required for the book projects and links to the product pages for the online retailers who sell these products. There are three different robot configurations used in the book, and this appendix covers the hardware needed for all of them. Make sure to consult this guide before buying the hardware. It is also worth noting that appendix D provides a mechanism to mock the robotic hardware and run all the code in the book on any laptop or desktop computer.

A.1 Raspberry Pi components

The Raspberry Pi is a small single-board computer created by the Raspberry Pi Foundation (https://raspberrypi.org). It is at the heart of all robot projects in this book. The Pi also supports an extensive set of add-on hardware boards that add additional functionality to the computer. We will need the following Raspberry Pi hardware for our projects:

- Raspberry Pi 4 (http://mng.bz/A8aK) with 2 GB or more of RAM
- Raspberry Pi Camera Module 2 (http://mng.bz/ZRVO), regular or NoIR version
- Adafruit camera case (https://www.adafruit.com/product/3253) for the Pi camera
- Adafruit CRICKIT HAT for Raspberry Pi (https://www.adafruit.com/product/3957) to control the DC and servo motors
- Pimoroni Pibow Coupe 4 (https://shop.pimoroni.com/products/pibow-coupe-4) case for the Raspberry Pi 4

There are a number of local and online retailers of these products around the world. Here are some helpful tips and sites to choose the best options for your location:

- The Raspberry Pi Foundation lists official retailers on each product page, which can be found when you click to buy a product on their website. The online tool lists the official retailers for a specific product and country.
- Adafruit products can be bought online (https://www.adafruit.com/) or through one of their official distributors (https://www.adafruit.com/distributors).
- Pimoroni products are available online (https://shop.pimoroni.com/) or through a distributor (http://mng.bz/RmN0).

The Raspberry Pi will need either a microSD card or a USB flash drive as storage. There are no specific storage requirements for the projects in the book as long as the space requirements to install the Raspberry Pi OS are met. Here are some points to keep in mind for different available storage options:

- If this is your first time working with a Raspberry Pi, then buying it as a kit provides a lot of extra items that are helpful for first-time users and are often good value for money. The kits will often come with a memory card for storage, a power supply, and an HDMI cable for video output. One such option is the Raspberry Pi 4 Desktop Kit (http://mng.bz/27gd). The Pimoroni Raspberry Pi 4 Essentials Kit (http://mng.bz/1JaV) is another popular option. This kit is also a good option if you find the regular Raspberry Pi 4 to be out of stock.
- USB flash drives can be significantly faster than microSD cards on the Raspberry Pi, which makes installing and upgrading software on the computer much faster as well. This article on Raspberry Pi storage performance gives more details on disk benchmarks and fast USB flash drives (http://mng.bz/PRN9).
- USB flash drives tend to be easier to change than microSD cards on the Raspberry Pi because of their location on the board. This is particularly true when the Raspberry Pi is fully assembled into a robotic chassis, as the USB ports are easier to access than the microSD slot. You might want to remove the storage so that you can easily take a full backup of the system on another computer, or you might have multiple USB flash drives, each with a different software setup, that you can swap out.

A.2 Motors, chassis kits, and joystick controllers

The two most common types of motors are DC and servo motors. Both types are used in this book. A robot chassis is also needed to attach the computer, motors, and batteries to. The recommended chassis kit has three layers, which gives more room for the board and battery than the smaller two-layer chassis kits:

- Adafruit Three-Layer Robot Chassis Kit (https://www.adafruit.com/product/3244) has two DC motors and wheels included.
- Adafruit Mini Pan-Tilt Kit (https://www.adafruit.com/product/1967) comes assembled with two micro servos that perform the pan and tilt movements.

Both kits are quite versatile and support many different hardware platforms. Their dimensions, connectivity, and power requirements are perfect for the Raspberry Pi using a CRICKIT HAT.

For chapter 7, which covers controlling robots with joysticks, there are a number of hardware options for the controller. An original Sony PlayStation 4 or 5 controller can be used. An Xbox original or compatible controller can be used as well. The following two Xbox-compatible controllers have been tested and work both with Linux and on the Raspberry Pi:

- W&O wireless controller compatible with Xbox 360 (https://a.co/d/7FA95aj)
- YAEYE controller for Xbox 360 (https://a.co/d/8lsabwI)

One thing to make note of is that wireless Bluetooth connectivity only works for the PlayStation controllers. You can, however, control the robot using a wireless network connection with any of the controllers using the approach of a Wi-Fi network connection to remotely control the robot, which is an approach covered in chapter 7.

A.3 Power and cabling

The Raspberry Pi and the CRICKIT HAT each need a power supply. There are many power options ranging from battery packs to connecting power cables to power outlets. The recommended approach is to use a single USB power bank to power both devices. There are a number of power banks that allow two devices to be connected and powered simultaneously. Power banks are rechargeable and portable. We need a portable power source to have our robots drive around without attached wires. Any USB power bank that supports simultaneously charging two devices can be used. The following power bank has been tested and works well:

- Anker PowerCore Select 10000 (https://walmart.com/ip/Anker/211593977) with Dual 12W output ports and 10000 mAh of power

The CRICKIT HAT receives power using a barrel jack connecter, so a USB to barrel jack cable is used to connect to the power bank. We also need extension jumper wires, as the cables that are part of the robot chassis kit will not be long enough to connect to the CRICKIT HAT once we have assembled all the required parts. The following are the recommended cables:

- USB to barrel jack cable (https://www.adafruit.com/product/2697)
- Premium M/M extension jumper wires (https://www.adafruit.com/product/1956)

A.4 Optional purchases

There are a number of items you can use to improve your robot-building experience, but these are not required. You will often want to take apart and reconfigure your robot hardware in different arrangements. You might be trying a different layout for your motors or modifying the position of your batteries to change the center of gravity of your robot. As you do this, you will want to be able to easily attach and detach your

Raspberry Pi and power bank from the chassis. Velcro adhesive squares are a great solution to this problem. When working with the Raspberry Pi, CRICKIT HAT, and robot chassis, there are a number of spots on each board and along the chassis to firmly screw parts together. The nylon screw and stand-off set provides many different screws and standoffs of different lengths for this exact purpose. Magnetic USB cables offer a clean way to easily connect and disconnect the power bank to and from the Raspberry Pi and to connect the power bank to a USB charger. The SlimRun Ethernet cables are lighter and thinner than standard network cables, which gives the robot more maneuverability when using a wired network connection:

- Velcro adhesive squares (https://a.co/d/8Sz6OMi)
- Nylon screw and stand-off set (https://www.adafruit.com/product/3658)
- Seven-pack magnetic USB cables (https://a.co/d/cYc3waP)
- Monoprice SlimRun Ethernet cable (https://a.co/d/0GBLsyQ)

appendix B
Configuring the Raspberry Pi

This appendix covers the installation and configuration of the main pieces of software on the Raspberry Pi. First, the Raspberry Pi OS Linux distribution will be installed on the computer. Then, Python will be configured so as to have a dedicated virtual environment where Python libraries can be installed. The Adafruit CRICKIT library will be installed next, which will then be used to run Python code to interact with the CRICKIT hardware.

B.1 Setting up the Raspberry Pi

The Raspberry Pi official documentation (https://raspberrypi.com/documentation/) page is an excellent source when working with the Raspberry Pi. The following sections of the documentation are good to look at for the projects in the book:

- Getting started: This page has detailed information on installing the operating system and using Raspberry Pi Imager.
- Configuration: Details on using the `raspi-config` tool can be found here.
- Remote access: It covers connecting to your Pi with SSH, transferring files, and using the VNC software to remotely access the desktop.

To install the Raspberry Pi OS, do the following:

1. Visit http://mng.bz/JdN0.
2. Click on Raspberry Pi OS link and download the "Raspberry Pi OS with desktop" image. This will download the latest release. For reference, the code in the book was tested with the 2022-04-04 32-bit release of Raspberry Pi OS.

The desktop image comes with a desktop environment, which will be useful when we create graphical applications for the robot projects.

3. Click on the "Raspberry Pi Imager" link and follow the instructions for downloading and installing the Imager software.
4. The Raspberry Pi 4 can be installed and booted from either a microSD card or a USB flash drive. USB flash drives offer better performance and are the recommended option.
5. Use the Imager software to prepare the installation media with the downloaded image (microSD card/USB flash drive).
6. Once the Raspberry Pi has booted with the installer, click Next at the welcome screen.
7. Set the value for country, and then set the username as `robo`, and continue with the configuration steps.
8. After reboot, we will use the Raspberry Pi configuration tool to configure the Pi further.
9. Use the tool to set the hostname to `robopi`. Figure B.1 shows the screen used to change the hostname on the Raspberry Pi.

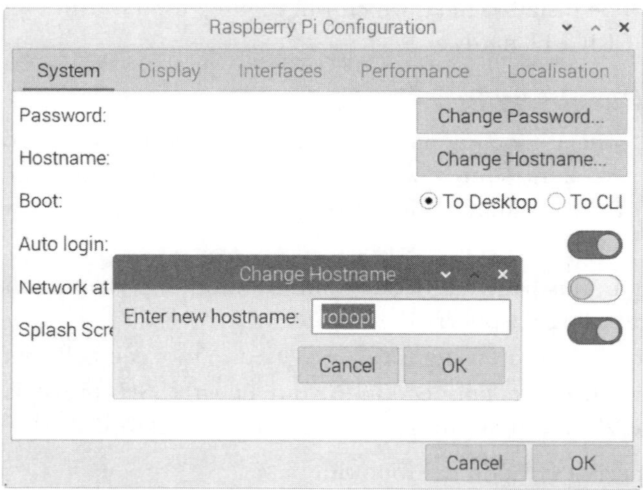

Figure B.1 Change hostname: use the Raspberry Pi configuration tool to change the hostname.

10. Next, use the configuration tool to enable the SSH, VNC, and I2C interfaces. Figure B.2 shows what the interfaces screen will look like once we have enabled these interfaces.

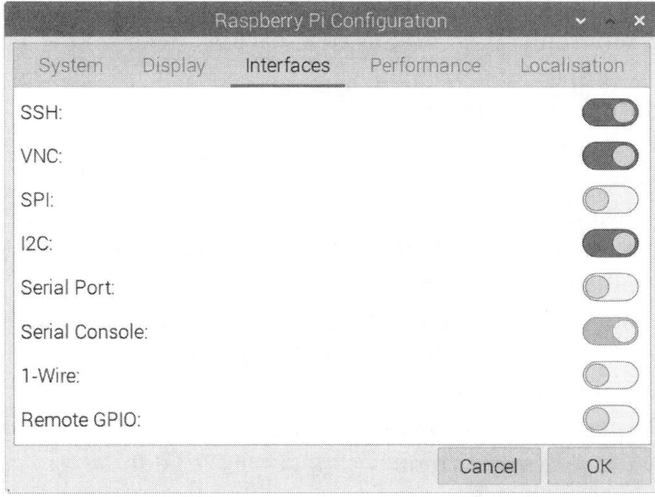

Figure B.2 Enabling interfaces: This screen can be used to enable different interfaces.

11. Now reboot the Pi for the changes to take effect.
12. Connect the Raspberry Pi to the network either by connecting a network cable to the Ethernet port or joining a Wi-fi network.
13. Get the IP address of the machine by running `hostname -I` in the terminal.
14. From another computer on the network, test that you can SSH to the Raspberry Pi using its IP address as the user `robo`. You can now use SSH to run commands and execute Python scripts on the Pi from any computer on your network.
15. You can also connect to the Raspberry Pi using its hostname `robopi`. To do this, you will need to add a line to your client machine's hosts file with the `robopi` hostname and its associated IP address. The How-To Geek website provides an excellent guide on how to edit the hosts file on Windows, Mac, and Linux (http://mng.bz/5owz).
16. The `sftp` command or the FileZilla application are both popular choices to transfer files to and from the Pi over the network. If your computer is running Linux, then `sshfs` is an excellent way to mount and work with remote files on the Pi as if they were local files.

Now that we have the main configuration steps completed for the Raspberry Pi, we can move on to configuring the Adafruit CRICKIT HAT.

B.2 Setting up the Adafruit CRICKIT HAT

Follow these steps to complete the hardware and software configuration of the Adafruit CRICKIT HAT:

1 There is a very comprehensive guide on the Adafruit website for setting up the CRICKIT HAT and troubleshooting any issues. We will refer to specific sections in the next steps (http://mng.bz/wj5q).

2 Before using the CRICKIT HAT for the first time, it is best to update its firmware. In the Adafruit learning guide, follow the steps in the "Update Your CRICKIT" section.

3 Switch off the Raspberry Pi. To attach the CRICKIT HAT to the Raspberry Pi, first connect the header standoff that comes with the CRICKIT to the Raspberry Pi GPIO connector. Then connect the CRICKIT HAT.

4 Connect the power cable into the CRICKIT DC jack, and switch on the CRICKIT power switch. Check whether the CRICKIT LED is green, which indicates there is a healthy power supply.

5 Power on the Raspberry Pi, and open a terminal or open an SSH connection to it.

6 Run the `i2cdetect` command, and check that the i2c address `0x49` appears in the output. The address will appear as the text `49` as shown:

```
$ i2cdetect -y 1
     0  1  2  3  4  5  6  7  8  9  a  b  c  d  e  f
00:                         -- -- -- -- -- -- -- --
10: -- -- -- -- -- -- -- -- -- -- -- -- -- -- -- --
20: -- -- -- -- -- -- -- -- -- -- -- -- -- -- -- --
30: -- -- -- -- -- -- -- -- -- -- -- -- -- -- -- --
40: -- -- -- -- -- -- -- -- -- 49 -- -- -- -- -- --
50: -- -- -- -- -- -- -- -- -- -- -- -- -- -- -- --
60: -- -- -- -- -- -- -- -- -- -- -- -- -- -- -- --
70: -- -- -- -- -- -- -- --
```

7 Run the following commands to update the software packages:

```
$ sudo apt update
$ sudo apt upgrade
$ sudo apt autoremove
```

8 Run the following command to reboot the machine:

```
$ sudo reboot
```

9 After rebooting, reconnect to the machine, and run the following lines to create a Python virtual environment and install the Adafruit CRICKIT library in that virtual environment:

```
$ python3 -m venv ~/pyenv
$ ~/pyenv/bin/pip install adafruit-circuitpython-crickit
```

10 Run the next line to add the `activate` Bash alias that can be used to activate the Python virtual environment whenever needed. After running the command, open a new terminal for the new alias to take effect:

```
$ echo "alias activate='source ~/pyenv/bin/activate'" >> ~/.bashrc
```

11 The next command will start a Python read–evaluate–print loop (REPL) session in the virtual environment:

```
$ ~/pyenv/bin/python
```

12 Run the following Python code in the REPL, and confirm that the on-board Neopixel turns red to configure the Adafruit CRICKIT HAT:

```
>>> from adafruit_crickit import crickit
>>> crickit.onboard_pixel.fill(0xFF0000)
```

B.3 Activating the Python virtual environment

We have now completed the setup and created a Python virtual environment. These virtual environments are a great way to keep our set of installed Python packages and environment separate from the system Python environment used by the operating system. In this way, we can recreate our robot's Python environment anytime we want without impacting the system installation. We can also install whatever packages and versions of them we want without worrying about breaking the Python packages the operating system is using. For more details on Python virtual environments, check the official documentation (https://docs.python.org/3/library/venv.html), which is a great resource on the topic.

There are some common operations with virtual environments that we will cover in this section. When you open a terminal or SSH to the Pi, you will get a prompt that looks like

```
robo@robopi:~ $
```

At this point, we have not activated our virtual environment. We can ask the terminal the location of the Python interpreter it would use with the following command:

```
robo@robopi:~ $ which python
/usr/bin/python
```

The output shows the location of the Python interpreter used by the operating system. Now run the `activate` command to activate our virtual environment:

```
robo@robopi:~ $ activate
(pyenv) robo@robopi:~ $
```

We can see that the text `(pyenv)` appears at the start of the prompt to indicate that we are in our Python virtual environment created in the previous section and called `pyenv`. We can now check the location of the Python interpreter again with the `which` command:

```
(pyenv) robo@robopi:~ $ which python
/home/robo/pyenv/bin/python
```

We can see whether it is using the Python interpreter in the virtual environment we have created for our robot projects. Now, we can open a REPL session in our virtual environment with the following command:

```
(pyenv) robo@robopi:~ $ python
Python 3.9.2 (default, Mar 12 2021, 04:06:34)
[GCC 10.2.1 20210110] on linux
Type "help", "copyright", "credits" or "license" for more information.
>>>
```

Press Ctrl+D to exit the REPL session. Now, when we execute Python scripts, they will run in our virtual environment and will be able to use the Python packages that we have installed.

appendix C
Robot assembly guide

This appendix covers how to assemble different hardware components into a completed robot. There are three different robot configurations used in the book, and this guide will cover how to build each. Make sure to go through appendixes A and B before using this guide. Each set of chapters uses a specific robot configuration:

- Chapters 2 through 7 use the mobile robot configuration.
- Chapters 8 and 9 use the servo camera robot configuration.
- Chapters 10 and 11 use the pusher robot configuration.

C.1 Building a mobile robot

The mobile robot configuration creates a robot that can drive around with an on-board power supply and can be controlled over a wireless connection. Follow these steps to build the robot:

1. The Adafruit website has an excellent guide on the CRICKIT HAT (http://mng.bz/qj0w). Follow the instructions in the guide to connect the CRICKIT HAT to the Raspberry Pi.
2. Then follow the Pibow Guide (http://mng.bz/7v59) to assemble and place the Raspberry Pi 4 in the Pibow case.
3. If you have purchased the optional nylon stand-off set, you can now place a stand-off on each of the two corners of the CRICKIT HAT that have no support to make the HAT sit on the Raspberry Pi in a more secure fashion.
4. Finally, connect the extension jumper wires to motor connections 1 and 2.
5. Figure C.1 shows what the Raspberry Pi looks like once these steps have been completed.

Figure C.1 Raspberry Pi: the Raspberry Pi and CRICKIT HAT are placed in the Pibow case.

The next set of steps involves preparing the power bank:

1. The dimensions, button layout, and port locations of different power banks will vary. However, fundamentally, the steps will be the same, although you might need to adjust the power bank placement.
2. The Anker PowerCore Select 10000 power bank has a button and charge indicator on the top part. We will place Velcro adhesive squares on the bottom of the power bank. These squares will help to attach the extra portions of cable under the power bank for cleaner cable management. Place adhesive squares on the power bank as shown in figure C.2.

Figure C.2 Power bank with adhesive squares: the squares are attached to the bottom of the power bank.

We will now proceed to connecting the power cables to the power bank:

1. Attach a magnetic USB cable to one of the power bank power output ports. This cable will be used to power the Raspberry Pi. The magnetic connector can be disconnected when the robot is not in use to reduce power drain on the power bank.

2. Plug the USB to Barrel Jack Cable into the other output port of the power bank. This cable will provide power to the CRICKIT HAT. Unlike the Raspberry Pi, the CRICKIT HAT has a power switch built into the board that can be used to disconnect the power supply, so it does not need to be unplugged when not in use.
3. Connect a magnet tip to the power bank power input port. This provides a convenient way to connect the power bank into a power supply to recharge the battery.
4. Figure C.3 shows what the power bank looks like once we have connected all the power cables.

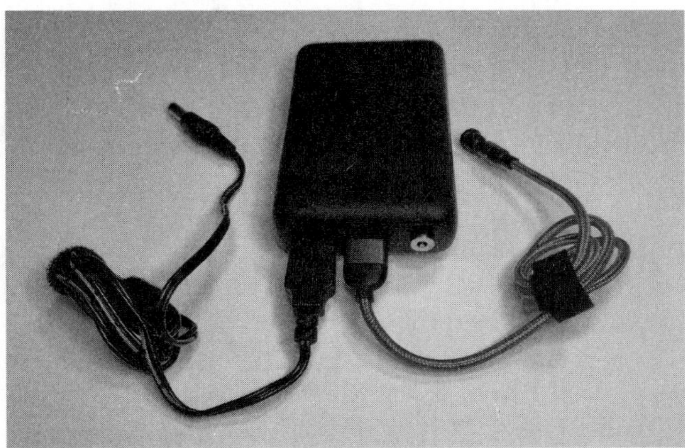

Figure C.3 Power bank with cabling: the USB cable is used to connect power to the Raspberry Pi.

We can now assemble the robot chassis and attach the power bank:

1. Follow the guide (http://mng.bz/mjRr) for the robot chassis kit to assemble all the parts.
2. The chassis kit is very flexible in that the plates on the chassis support many different arrangements for the screws and brass standoffs. You can experiment with different arrangements to see what best suits the specific dimensions of your power bank. The power bank can now be placed in the middle layer of the chassis.
3. It is best to keep the power bank in the middle layer instead of the top layer when possible. It is the heaviest part of the robot, and the robot will be less likely to tip over when the weight is kept in a lower layer. This is most noticeable when the robot accelerates to the maximum from a resting position to top speed or when it decelerates at a high rate to come to a full stop.
4. Adhesive squares can now be attached to the top layer of the chassis. Four adhesive squares are attached to provide a very firm attachment between the Pibow case and the chassis.

5. Figure C.4 shows what the robot chassis looks like with the power bank in place.

Figure C.4 Chassis with power bank: the Raspberry Pi is placed on the top layer of the chassis.

The final part of the assembly is attaching the Raspberry Pi to the chassis:

1. Attach the Raspberry Pi with its case to the top layer of the chassis. The USB ports on the Raspberry Pi should be facing the back of the robot. This keeps the power connectors for the CRICKIT HAT and Raspberry Pi closer to the power bank power cables.
2. Connect the power cables for the CRICKIT HAT and Raspberry Pi.
3. Finally, connect the jumper wires to the DC motors. Make sure to connect the right DC motor to motor connection 1 and the left DC motor to motor connection 2.
4. Figure C.5 shows a fully assembled robot.

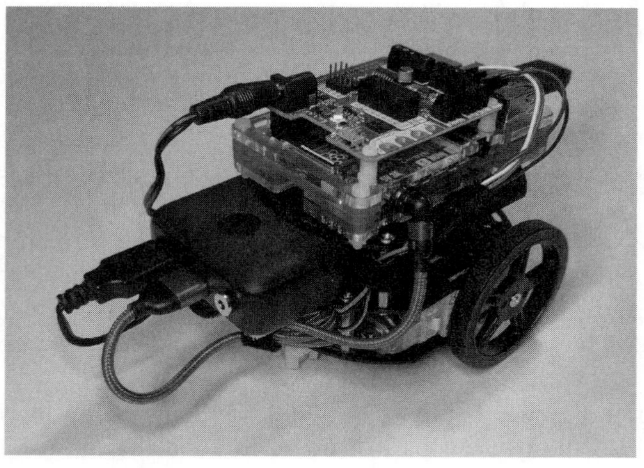

Figure C.5 Fully assembled robot: the Raspberry Pi USB ports are accessible from the rear of the robot.

C.2 Building a servo camera robot

The servo camera robot configuration creates a robot with a camera attached to two servo motors. One servo will allow us to pan the camera, and the other will apply a tilt motion. Follow these steps to build the robot:

1 The Adafruit Mini Pan-Tilt Kit comes fully assembled. The kit supports mounting different sizes and styles of camera modules. This Adafruit robot assembly guide (http://mng.bz/5oOB) uses the Pan-Tilt Kit and has a good explanation of removing the side tabs that will be needed for the next step.
2 The kit has two side tabs made of soft nylon that can be removed. Remove one of the side tabs either by twisting it off or by using a wire cutter. We only need one tab to be in place.
3 Place a Velcro adhesive square on the kit camera mounting point.
4 The base of the kit should be attached to a sturdy surface for better stability. You can cut out a piece of cardboard and attach the kit to the cardboard using glue or double-sided foam tape.
5 Figure C.6 shows what the Pan-Tilt Kit will look like at this point.

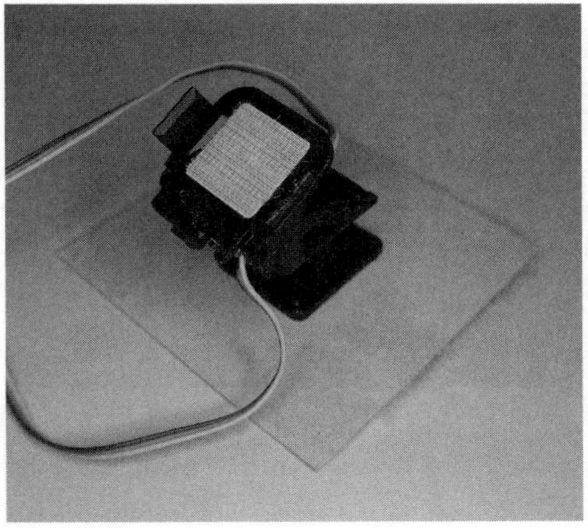

Figure C.6 Pan-Tilt Kit with Velcro: the Velcro square will be used to attach the camera.

The next set of steps involves preparing the camera:

1 The Raspberry Pi Camera Module can now be placed inside the Adafruit camera case.
2 The camera case is very well suited for mounting to the kit, as it has slots on either side of the camera that fit well into the kit side tabs.
3 Place a Velcro adhesive square on the back of the camera case.
4 Figure C.7 shows what the camera case should look like with the Velcro squares.

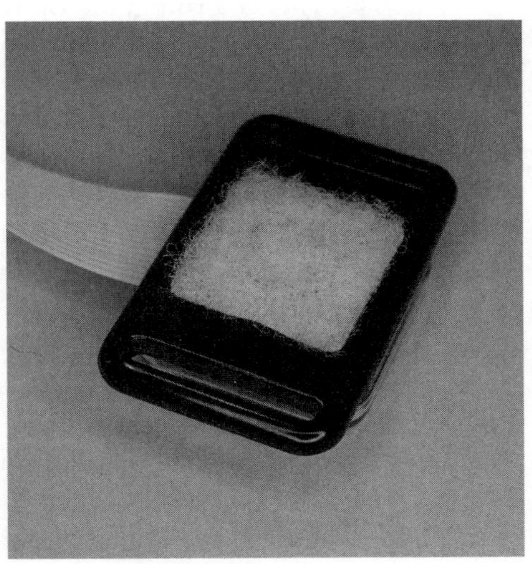

Figure C.7 Camera case: the Velcro squares are attached to the back of the camera case.

We will now proceed to attach the camera case to the Pan-Tilt Kit:

1 Using Velcro squares will let us attach and detach the camera as needed.
2 You can attach the camera case to the Pan-Tilt Kit now. Make sure the camera cable is at the bottom.

3. This is the correct orientation of the camera so that images taken from the camera will be right side up.
4. The side tab on the Pan-Tilt Kit will slide into the slot in the camera case.
5. Figure C.8 shows what the camera and Pan-Tilt Kit will look like once attached.

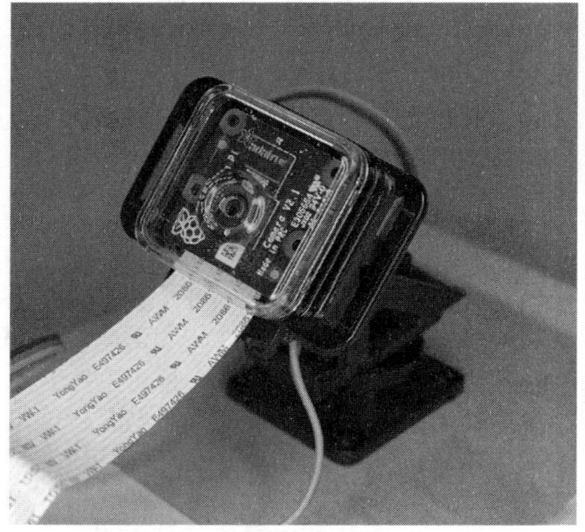

Figure C.8 Camera attached to the Pan-Tilt Kit: the camera will be attached to the kit using Velcro.

The final steps of the assembly are to connect the Pan-Tilt Kit to the Raspberry Pi:

1. The official documentation on the Raspberry Pi Camera Module (http://mng.bz/6nYo) offers excellent information on how to connect the camera cable to the Raspberry Pi.
2. The CRICKIT HAT has an opening for the camera cable. Pass the camera cable through the CRICKIT HAT and connect it to the Raspberry Pi. Then attach the CRICKIT HAT to the Raspberry Pi.
3. Now place the Raspberry Pi 4 in the Pibow case.
4. The Adafruit guide on the CRICKIT HAT referred to in the last section is an excellent resource to look at for connecting the Pan-Tilt Kit to the CRICKIT HAT. Specifically, the CircuitPython Servos section provides good details on how to connect the servo motors on the Pan-Tilt Kit to the CRICKIT HAT.
5. The guide explains the orientation in which to attach the servo connectors to the CRICKIT HAT. The wires on one side of the servo connector will have dark colors, such as black and brown. The other side will use lighter color wires, such as yellow, orange, or white. Connect the dark wires toward the CRICKIT logo and the light colors toward the DC power jack.
6. Connect the lower servo connector to servo connection 1 and the upper servo connector to servo connection 2.
7. Now connect the power cables for the CRICKIT HAT and Raspberry Pi like we did in the previous section.
8. Figure C.9 shows what the fully assembled robot will look like. This robot can move the attached camera in both the pan and tilt directions using the servo motors. It does not make use of the DC motors like the previous robot. The robot configuration in the next section makes use of both the servo and DC motors.

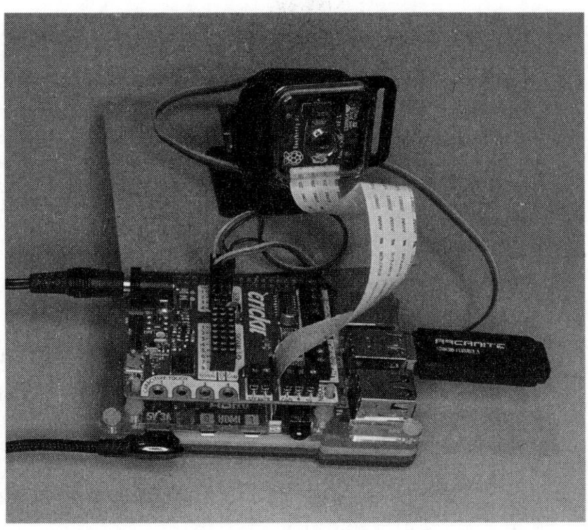

Figure C.9 Servo camera robot: the camera can move around using the servo motors.

C.3 Building a pusher robot

The pusher robot configuration creates a robot that can drive back and forth along a track using the DC motors, and then, using the camera, will look for objects with a matching QR code. Once found, the items can be pushed off a counter using an arm attached to a servo motor. This robot combines the two previous robot configurations in many ways. This configuration will take the mobile robot configuration and add a camera and servo motor to it. Follow these steps to build the robot:

1. Complete the mobile robot configuration. Then detach the Pibow case from the robot chassis.
2. Cut out a piece of cardboard, attach the cardboard to the chassis, and then attach the Pan-Tilt Kit to the cardboard using glue or double-sided foam tape. Figure C.10 shows what this will look like from a top view. The strip of cardboard

Figure C.10 Pan-Tilt Kit attached to chassis: the photo shows the top view of the Pan-Tilt Kit and robot chassis.

is cut so that it can be placed between the Velcro squares and the power bank button. Alternatively, a short ruler can also be used instead of a cardboard cutout. Make sure to place the Pan-Tilt Kit on the side of the Raspberry Pi that has no ports. It is the side opposite of the one with the HDMI and USB power ports. This side has no ports, so putting a camera and servo on it won't block our access to any ports.

The next steps of the assembly are to attach the camera:

1. We can now reattach the Pibow case to the robot chassis by placing the Velcro squares on top of each other like before.
2. We then stick some Velcro squares to the side of the Pibow case so that we can attach the camera to the side of the robot. Figure C.11 shows what the side of the robot will look like once we attach these adhesive squares.
3. Next, we attach the camera to the side of the robot. Figure C.12 shows what the robot looks like once the camera is attached. The camera case will rest on the robot chassis once attached. The ribbon will be coming out at the top of the camera case. This means the camera will capture video upside down. This won't be a problem, as we can correct this in software by flipping the image during video capture.

Figure C.11 Camera Velcro squares: Velcro squares are attached to the side of the Pibow case.

Figure C.12 Camera attached: the camera is attached to the side of the robot.

The final step of the assembly is to create a servo arm for the robot:

1 We will use the tilt servo of the Pan-Tilt Kit as the servo to push detected objects off the counter. We want to attach an arm to extend the physical range of our tilt servo. The mounting bracket on the kit has a side tab and slots we can use to secure an arm on the tilt servo.
2 We can use a pencil for the arm, as it is sturdy and light enough with an appropriate length to provide our arm some good pushing range. Position the eraser at the bottom of the arm, as it will provide a softer surface when it makes contact with the objects we are pushing. Attach the pencil using two zip ties: one zip tie above the side tab and one below it. Figure C.13 shows the side view of the robot with the servo arm attached.
3 Figure C.14 shows a rear view of the robot. From this view, you can get a better view of how one zip tie is placed above the side tab and one below it. This will hold the arm firmly in place while it is repeatedly raised and dropped. Make sure to tighten the zip ties so that they can have a firm hold on the servo arm.

Figure C.13 Servo arm side view: the photo shows the side view of the robot with the servo arm attached.

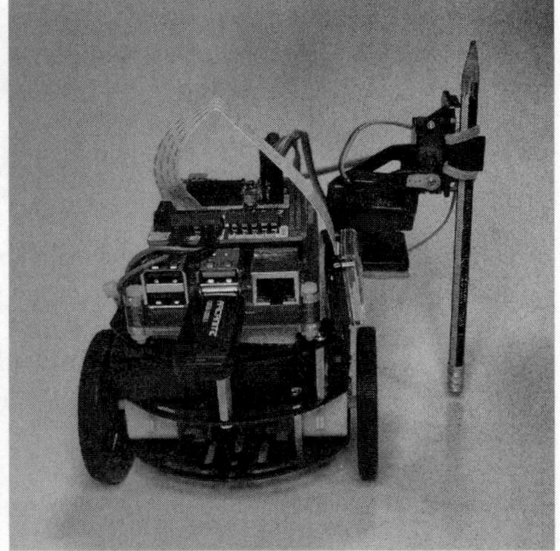

Figure C.14 Servo arm rear view: the photo shows the rear view of the robot with the servo arm attached.

The robot is now completed and ready to be placed on the track. Check the next section for further details on creating a track for the robot.

C.4 Creating a track for the pusher robot

The pusher robot drives backward and forward along a track. This is done in the same way that the path of a train is controlled by restricting its movements to go back and forth along the path of the train track. We can use two sticks or poles to create a track for the robot. We'll put some books at each of the track ends to make sure the track doesn't shift around as the robot moves along the track. Follow these steps to create the track:

1. Take any two sticks or poles and place them parallel to each other on a table. Figure C.15 shows two sticks placed in parallel on our table. The sticks in the photo are two broomsticks with the brush screwed off. Any sticks or poles can be used.
2. Keep the distance between the sticks as close as possible so that they are hugging the tires of the robot. Figure C.15 also shows how the items the robot will push can be placed along the edge of the table with their QR codes pointing toward the robot so that it can read their QR codes as it drives down the track.

Figure C.15 Robot track: the track should hug the robot tires.

3. Figure C.16 shows a set of books that have been put on each end of the track. They will keep the track firmly in place as the robot moves back and forth.

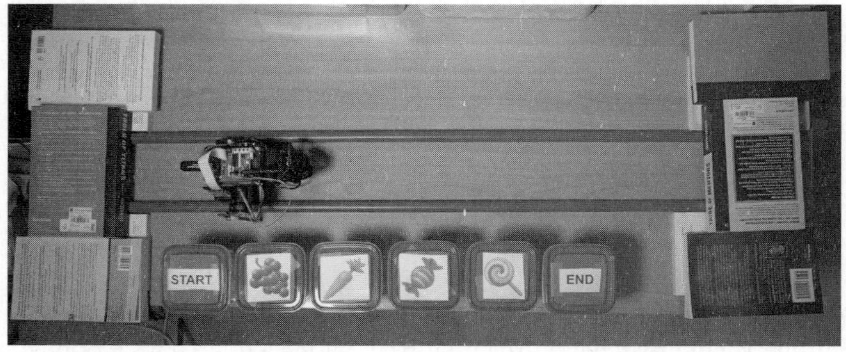

Figure C.16 Track with books: we use books to keep the track firmly in place.

4 The start marker should be placed at the start of the track, and the end marker should be placed as the last item along the track. Figure C.15 shows the start and end markers, as well as four snacks positioned between them.

The last part of the setup is the snack boxes:

1 Figure C.17 is a photo of one of the snack boxes. These can be any container or packet with our desired QR code placed on the front of the container. The QR codes on the label should be large enough to be easily read by the camera. A width and height of 6 cm for the QR codes were tested and work well.
2 The QR code should face the robot so that it can see it when it drives by. During testing, a distance of 8 cm between the camera and QR codes worked well.

Figure C.17 Snack box: the snack box has a QR code in the front.

appendix D Mocking the CRICKIT library

This appendix covers the topic of mocking the Adafruit Python CRICKIT library. Mocking is a common mechanism used when testing software. It lets us replace a piece of software with mock objects. In our case, there are a number of benefits to doing this:

- *Running code without robot hardware*—The mock library in this appendix will let us run all the code in the book without any robot hardware. This is useful, as it provides a deeper look at the computer vision, web, joystick, and networking portions of the code that don't need robotic hardware.
- *Executing the code without a Raspberry Pi*—The code is written to run on many different Linux systems. All the code was tested with Ubuntu 22.04, which can be run on any laptop or virtual machine under Windows or Mac.
- *Better coding experience*—Running on a modern computer can often be a faster and more comfortable development experience than executing the code on slower machines such as the Raspberry Pi. For example, when doing a lot of work on the web front end of a robot application, the development cycle can be faster and more productive on a laptop.

D.1 Installing the mock CRICKIT library

The mock object library (http://mng.bz/or0d) that comes as part of the Python standard library will be used to mock the different functions of the CRICKIT library. The implementation will address mocking the specific functionality used by the code in this book and not the whole CRICKIT library. This appendix will mainly focus on using the `mock_crickit` library and will not cover the implementation details in great depth. Save the following script in a file called `adafruit_crickit.py`.

Listing D.1 `adafruit_crickit.py`: Mocking the CRICKIT library

```
#!/usr/bin/env python3
import os
from unittest.mock import Mock, PropertyMock
from functools import partial

DEBUG = bool(os.environ.get('ROBO_DEBUG'))
PROP_VALUES = {'touch_1.value': True}

def print_msg(msg):
    if DEBUG:
        print('MOCK_CRICKIT:', msg)

def prop_access(name, *args):
    action = 'set' if args else 'get'
    if action == 'set':
        PROP_VALUES[name] = args[0]
    val = PROP_VALUES.get(name)
    print_msg(f'{action} crickit.{name}: {val!r}')
    return val

def pixel_fill(val):
    print_msg(f'call crickit.onboard_pixel.fill({val!r})')

def add_property(name):
    parent, child = name.split('.')
    property_mock = PropertyMock(side_effect=partial(prop_access, name))
    setattr(type(getattr(crickit, parent)), child, property_mock)

crickit = Mock()
crickit.onboard_pixel.fill = Mock(side_effect=pixel_fill)
names = [
    'onboard_pixel.brightness', 'touch_1.value', 'dc_motor_1.throttle',
    'dc_motor_2.throttle', 'servo_1.angle', 'servo_2.angle',
    'servo_1.actuation_range', 'servo_2.actuation_range']
for name in names:
    add_property(name)

def demo():
    print('starting mock_crickit demo...')
    crickit.onboard_pixel.brightness = 0.01
    crickit.onboard_pixel.fill(0xFF0000)
    crickit.touch_1.value
    crickit.dc_motor_1.throttle = 1
    crickit.dc_motor_2.throttle = -1
    crickit.servo_1.angle = 70
    crickit.servo_1.angle
    crickit.servo_2.angle = 90
    crickit.servo_2.angle
    crickit.servo_1.actuation_range = 142
    crickit.servo_2.actuation_range = 180

if __name__ == "__main__":
    demo()
```

The library is written as a direct replacement for the `adafruit_crickit` library, which is why it has the same name. We can use it in place of the Adafruit library without changing our Python code. As we have done throughout the book, we can set the `ROBO_DEBUG` environment variable to have the mock library print out every mock call it receives. When the library is directly executed, it will perform the `demo` function that demonstrates all the different parts of the CRICKIT library that it mocks. The following session shows a sample run of the library:

```
$ export ROBO_DEBUG=1
$ ./adafruit_crickit.py
starting mock_crickit demo...
MOCK_CRICKIT: set crickit.onboard_pixel.brightness: 0.01
MOCK_CRICKIT: call crickit.onboard_pixel.fill(16711680)
MOCK_CRICKIT: get crickit.touch_1.value: True
MOCK_CRICKIT: set crickit.dc_motor_1.throttle: 1
MOCK_CRICKIT: set crickit.dc_motor_2.throttle: -1
MOCK_CRICKIT: set crickit.servo_1.angle: 70
MOCK_CRICKIT: get crickit.servo_1.angle: 70
MOCK_CRICKIT: set crickit.servo_2.angle: 90
MOCK_CRICKIT: get crickit.servo_2.angle: 90
MOCK_CRICKIT: set crickit.servo_1.actuation_range: 142
MOCK_CRICKIT: set crickit.servo_2.actuation_range: 180
```

We can also install the mock library into any Python virtual environment of our choosing. The code for the library and the installer can be found on GitHub (https://github.com/marwano/robo). In the next session, we will install the `mock_crickit` library using the `pip install` command. Make sure to run the `pip install` command in the same directory as the setup script:

```
(main) robo@robopi:/tmp$ cd mock_crickit
(main) robo@robopi:/tmp/mock_crickit$ pip install .
Processing /tmp/mock_crickit
 Installing build dependencies ... done
 Getting requirements to build wheel ... done
 Preparing metadata (pyproject.toml) ... done
Building wheels for collected packages: mock-crickit
 Building wheel for mock-crickit (pyproject.toml) ... done
 Created wheel for mock-crickit: filename=mock_crickit-1.0-py3-none-any.whl
Successfully built mock-crickit
Installing collected packages: mock-crickit
Successfully installed mock-crickit-1.0
```

We can now run `pip list` to obtain a list of installed packages in our virtual environment. We can see that we have installed version `1.0` of the `mock-crickit` library:

```
(main) robo@robopi:/tmp/mock_crickit$ pip list
Package      Version
------------ -------
mock-crickit 1.0
pip          23.1.2
setuptools   59.6.0
```

We can verify the library is functioning correctly by calling the `demo` function with the following command:

```
(main) robo@robopi:~$ python -m adafruit_crickit
starting mock_crickit demo...
MOCK_CRICKIT: set crickit.onboard_pixel.brightness: 0.01
MOCK_CRICKIT: call crickit.onboard_pixel.fill(16711680)
MOCK_CRICKIT: get crickit.touch_1.value: True
MOCK_CRICKIT: set crickit.dc_motor_1.throttle: 1
MOCK_CRICKIT: set crickit.dc_motor_2.throttle: -1
MOCK_CRICKIT: set crickit.servo_1.angle: 70
MOCK_CRICKIT: get crickit.servo_1.angle: 70
MOCK_CRICKIT: set crickit.servo_2.angle: 90
MOCK_CRICKIT: get crickit.servo_2.angle: 90
MOCK_CRICKIT: set crickit.servo_1.actuation_range: 142
MOCK_CRICKIT: set crickit.servo_2.actuation_range: 180
```

The projects in the book can be executed with this library on a wide variety of hardware. The joystick hardware mentioned in appendix A can be used with any laptop or desktop computer running Linux. Furthermore, any standard webcam can be used in place of the Raspberry Pi Camera Module with no modification to the code in the book. This enables the computer vision functionality of face and QR code detection.

index

Symbols

Numerics

A

B

C

D

N

O

P

S

T

U

V

W

X